Wang Shaoqiang (ed.)

PACKAGING
MEETS
CREATIVITY

LE PACKAGING
CRÉATIF

PACKAGING
CREATIVO

UNPACK
ME AGAIN!

promopress

UNPACK ME AGAIN!

PACKAGING MEETS CREATIVITY
LE PACKAGING CRÉATIF
PACKAGING CREATIVO

Editor: Wang Shaoqiang
English preface revised by: Tom Corkett
Translators of the preface:
Leïla Bendifallah, French translation
Jesús de Cos Pinto, Spanish translation

Promopress is a brand of:
Promotora de Prensa Internacional S.A.
C/ Ausiàs March, 124
08013 Barcelona, Spain
Phone: 0034 93 245 14 64
Fax: 0034 93 265 48 83
info@promopress.es
www.promopresseditions.com
Facebook: Promopress Editions
Twitter: Promopress Editions @PromopressEd
Sponsored by Design 360°
– Concept and Design Magazine
Edited and produced by
Sandu Publishing Co., Ltd.
Book design, concepts & art direction by
Sandu Publishing Co., Ltd.
info@sandupublishing.com

Cover design:
spread: David Lorente

ISBN 978-84-16504-53-4

Printed in China

CONTENTS

PACKAGING
MEETS
CREATIVITY

LE PACKAGING CRÉATIF
PACKAGING CREATIVO

THOUGHTS ON PACKAGING DESIGN

Joshua Breidenbach
Rice Creative

As a designer who has made a lot of packaging and spent countless hours striving to create functional design with meaning and soul, my thoughts on packaging tend not to be on design per se, but rather on the meaning of it all. I'm taking this opportunity to reflect and to pack up an agglomeration of thoughts on my process and personal terms for judging a successful outcome.

A packaging design assignment is an endlessly interesting endeavor for some. The process twists and turns, finally terminating with a compact, completely tangible result that speaks a language. That result might be reproduced thousands or millions of times and be seen by thousands or millions of people all over the world. It takes up significant space in the world, and it will be touched and scrutinized by masses of people. It is a message sent out into space and it will have some form of influence. The practice carries similarities to architecture and advertising, though the product more often finds itself actually living with people inside of millions of homes. I find these facts as heavy as I do exhilarating. They remind me of my responsibility as a designer, and I never forget how my work could impact our community culturally, ecologically, socially, and ethically.

To complete a thorough packaging design assignment, I must weigh several factors at once. Over time, I have developed a mental checklist to keep my process on the right path. First of all, functionality is of paramount importance. The package clearly has to hold and protect its contents. It is part of the product and adds great value. A phenomenal piece of packaging design definitely makes a dramatic impact on the sales and distribution of the product.

Packaging design is an act of telling the product's story. A package has to be meaningful. I can make a powerful tool to connect with an audience and convey my client's story well. It is a greeting from brand to individual. It is a single-image narrative like a painting. I am always aiming to make the package an icon. It should be as succinct as a logo. The package will

be traded across nations and among different cultures. Packaging is a thing that should be recognized universally if possible, in the same way that a very good poster would be.

At the same time, I aim equally for frugality. For me, the life span of a design is a significant concern. I find it imperative for a package to last beyond the trip home from the shop. I am seeking to be ecological for the Earth and economical for the client. I do not want to create rubbish; there is already so much of it. If at all possible, I make something that will easily have a second life.

At the very least, I hope that the thing that I have created will offer some form of inspiration. I want the package to cause a reaction, whether it is a thought, a raised eyebrow, or a smile. I want the package to invite curiosity and to encourage connections and conversations. I ensure that the entire endeavor is worthwhile to my client. The worst outcome of all is a design that fails the first goal: helping my client to achieve their goals. All of these aims, as lofty as some might seem, go hand in hand, helping an audience find affinity with the brand that I have agreed to support.

Seeing a finished packaging design in the market is a wonderful feeling, but it often provokes self-reflection. Questions flood my mind: Will it work for my client? Could the structure fail? Is it going to actually help sell? Have I convinced my client to do the right thing? Have I wasted their money when they could have invested elsewhere? The list can go on and on, but I find comfort in my list, because if I follow it, these doubts lead to answers, and I may be able to achieve something extra for both my client and myself.

We here at Rice are delighted to see our works in *Unpack Me Again: Packaging Meets Creativity*. I hope this meandering introduction, along with the beautifully crafted packaging projects featured in this book, sparks further thought on this field.

Joshua Breidenbach is a designer, a creative director, and the co-founder of Rice Creative, a Vietnam-based branding and creative agency. Rice Creative has gained an international reputation in little over five years for highly creative and effective work for a wide spectrum of clients. Joshua moved to Vietnam in 2006 as design director of Lowe & Partners. His ambition became set on creating world-class creative solutions for emerging brands, visionary companies, and causes in Asia. Joshua went on to co-found the multidisciplinary practice Rice Creative in 2011. Joshua has received accolades for his work from Red Dot, Pentawards, A'Design, Graphis, the Dieline, and Type Directors Club, as well as an IDA award for "Graphic Design of the Year" for his work with UNICEF Vietnam.

RÉFLEXIONS SUR LE DESIGN D'EMBALLAGE

Joshua Breidenbach
Rice Creative

En tant que designer ayant créé de nombreux emballages et ayant passé un nombre incalculable d'heures à faire mon possible pour créer un design fonctionnel doté de sens et d'une âme, mes réflexions sur l'emballage ont tendance à ne pas concerner le design en soi mais plutôt le sens de tout cela. Je profite de cette occasion pour réfléchir et rassembler une agglomération de pensées sur le processus et les conditions personnelles que j'applique pour juger d'un aboutissement réussi.

Pour certains, une mission de design d'emballage est un effort à l'intérêt infini. Le processus est sinueux et s'achève avec un résultat compact et totalement tangible qui parle une langue. Ce résultat peut être reproduit des milliers ou des millions de fois et être vu par des milliers ou des millions de personnes dans le monde entier. Il prend un espace important dans le monde et il sera touché et examiné par un grand nombre de personnes. C'est un message envoyé dans l'espace et qui aura une certaine forme d'influence. Cette pratique a des points communs avec l'architecture et la publicité, bien que le produit se retrouve plus souvent vivant vraiment avec les individus dans des millions de foyer. Je trouve cela aussi profond qu'exaltant. Cela me rappelle ma responsabilité en tant que designer et je n'oublie jamais l'impact culturel, écologique, social et éthique que mon travail peut avoir sur notre communauté.

Pour terminer une mission de design d'emballage minutieuse, je dois prendre en compte différents facteurs à la fois. Avec le temps, j'ai élaboré une check-list mentale me permettant de ne pas m'égarer dans mon processus. Tout d'abord, la fonctionnalité est d'une importante primordiale. Il est évident que l'emballage doit tenir et protéger son contenu. Il fait partie du produit et y ajoute une grande valeur. Il ne fait aucun doute qu'un excellent design d'emballage a un impact considérable sur la vente et la distribution d'un produit.

Le design d'emballage consiste à raconter l'histoire du produit. Un emballage doit avoir un sens. Je peux créer un outil puissant pour entrer en contact avec le public et bien communiquer l'histoire de mon client. C'est un message de la marque pour l'individu. C'est un récit à image unique, comme une peinture. Je vise toujours à faire de l'emballage une icône. Il devrait être aussi succinct qu'un logo. Il sera vendu à travers différents pays et parmi différentes cultures. Un emballage doit être reconnu de façon universelle si possible, à la manière d'une très bonne affiche.

Dans le même temps, je vise tout autant la sobriété. Pour moi, la durée de vie d'un design fait l'objet d'une préoccupation importante. Il est impératif à mes yeux qu'un emballage vive plus longtemps que ce que dure le trajet du magasin à la maison. Je cherche à être écologique pour la planète et économique pour le client. Je ne souhaite pas créer des déchets; il y en a déjà tellement. Si cela est possible, je crée des choses qui auront facilement une deuxième vie.

J'espère que ce que je crée apportera une quelconque forme d'inspiration. Je veux que l'emballage provoque une réaction, qu'il s'agisse d'une pensée, d'un haussement de sourcil ou d'un sourire. Je veux qu'il invite la curiosité et qu'il favorise les liens et les conversations. Je garantis que la totalité du travail vaut la peine pour mon client. Le pire résultat possible est un design n'atteignant pas le premier objectif: aider mon client à parvenir à ses fins. Tous ces objectifs, aussi ambitieux que certains puissent paraître, vont ensemble pour aider un public à trouver des affinités avec la marque que j'ai accepté de soutenir.

Voir un design d'emballage terminé sur le marché apporte un sentiment merveilleux, mais cela suscite souvent une introspection. Mon esprit est submergé par les questions: Cela fonctionnera-t-il pour mon client? La structure pourrait-elle échouer? Va-t-elle vraiment favoriser la vente? Ai-je convaincu mon client de faire ce qui est juste? Ai-je gaspillé son argent alors qu'il aurait pu investir autre part? La liste est interminable, mais elle me rassure car, si je la suis, ces questions mènent à des réponses et peuvent me permettre d'obtenir quelque chose en plus pour le client et moi-même.

Chez Rice, nous sommes ravis de voir nos travaux exposés dans *Unpack Me Again: Le packaging créatif*. J'espère que cette introduction nébuleuse, ainsi que les projets d'emballage magnifiquement réalisés figurant dans cet ouvrage, susciteront davantage de réflexion sur ce sujet.

Joshua Breidenbach est designer, directeur créatif et cofondateur de Rice Creative, une agence de création et de stratégie de marque basée au Vietnam. Rice Creative a acquis une réputation internationale en un peu plus de cinq ans grâce à son travail très créatif et efficace pour une grande variété de clients. Joshua s'est installé au Vietnam en 2006 en tant que directeur design de Lowe & Partners. Il a concentré son ambition sur l'idée de créer des solutions créatives d'envergure internationale pour des marques émergentes, des entreprises visionnaires et des causes en Asie. Joshua a ensuite cofondé l'agence multidisciplinaire Rice Creative en 2011. Il a reçu des récompenses pour son travail de la part de Red Dot, Pentawards, A'Design, Graphis, the Dieline et Type Directors Club, ainsi qu'un prix IDA du « Meilleur design graphique de l'année » pour son travail avec l'UNICEF Vietnam.

REFLEXIONES SOBRE EL DISEÑO DE PACKAGING

Joshua Breidenbach
Rice Creative

Como autor de gran cantidad de diseños de packaging que ha dedicado incontables horas a intentar crear diseños funcionales con significado y con alma, mis pensamientos sobre el packaging no se centran en el diseño en sí sino en su significado global. Aprovecho esta ocasión para reflexionar y reunir una serie de pensamientos sobre mis procesos y sobre mis criterios personales para evaluar la calidad de los resultados.

Para algunos, el encargo de diseñar un embalaje es una tarea de enorme interés. El proceso de trabajo es un camino sinuoso y fortuito que desemboca en un resultado compacto y tangible que habla un idioma propio. Este resultado será reproducido miles o millones de veces y visto por miles o millones de personas en todo el mundo. Ocupará un espacio significativo y será tocado y escrutado por las multitudes. Es un mensaje enviado al espacio y tiene algún tipo de influencia. Esta actividad tiene similitudes con la arquitectura y con la publicidad, y sus creaciones proliferan en la vida diaria de millones de hogares. Para mí, estas circunstancias son tan serias como estimulantes, porque me recuerdan mi responsabilidad como diseñador y me obligan a tener en cuenta el posible impacto cultural, ecológico, social y ético de mi trabajo.

Para cumplir un encargo de diseño debo sopesar diversos factores simultáneamente. Con el tiempo, he desarrollado una lista de comprobación mental para encauzar mi trabajo. Ante todo, la funcionalidad es de primordial importancia: el embalaje tiene que incluir y proteger su contenido. El embalaje es parte del producto y le añade mucho valor. Un trabajo de packaging espléndido tiene un impacto decisivo en las ventas y en la distribución del producto.

El diseño del packaging es el acto de narrar la historia del producto. El embalaje debe tener significado. Como diseñador, puedo crear una herramienta poderosa para comunicar con el público y trasladarle el mensaje de mi cliente. Es un saludo de la marca dirigido al individuo; es un relato en una sola imagen, igual que un cuadro. Siempre intento conseguir que sea un icono, algo tan sucinto como un logo. El embalaje viaja a través de naciones y de culturas y, si es posible, debe ser reconocible universalmente, lo mismo que un póster muy bueno.

Al mismo tiempo, aspiro a la frugalidad. Para mí, la vida útil de un diseño es una cuestión importante: me siento obligado a hacer que el embalaje dure más allá del viaje desde la tienda a casa. Deseo a ser ecológico para la Tierra y económico para el cliente. No quiero fabricar basura, ya hay demasiada. Dentro de lo posible, intento hacer algo que pueda tener una segunda vida.

Como mínimo, espero que el objeto que he creado ofrezca algún tipo de inspiración. Quiero que el embalaje cause una reacción, ya sea ésta un pensamiento, una ceja levantada o una sonrisa. Quiero que excite la curiosidad y fomente conexiones y conversaciones. Me aseguro de que mi labor sea valiosa para mi cliente. El peor resultado posible es el diseño que no logra el primer objetivo: ayudar a mi cliente a conseguir sus metas.

Todos estos propósitos, aunque algunos de ellos puedan parecer ambiciosos, van juntos de la mano y ayudan a que el público encuentre una afinidad con la marca que yo he aceptado defender.

Ver el packaging acabado y en el mercado es una sensación fantástica, pero a menudo me conduce al autoexamen. Las preguntas fluyen a mi mente: ¿será eficaz para mi cliente? ¿tendrá fallos estructurales? ¿contribuirá de verdad a las ventas? ¿he convencido a mi cliente de que haga lo correcto? ¿he malgastado un dinero que él podría haber invertido mejor? Las preguntas continúan indefinidamente, pero mi lista de comprobación me reconforta porque, si la sigo, mis dudas tienen respuestas y quizá seré capaz de conseguir algo especial tanto para mi cliente como para mí mismo.

En Rice estamos encantados de ver nuestros trabajos publicados en *Unpack Me Again: packaging creativo*. Espero que esta errática introducción, junto con los hermosos proyectos de packaging incluidos en este libro, inciten a nuevas reflexiones sobre el tema.

Joshua Breidenbach es diseñador, director creativo y cofundador de Rice Creative, una agencia de creatividad y creación de marca radicada en Vietnam. Rice Creative ha obtenido prestigio internacional en poco más de cinco años por sus trabajos creativos y eficaces para una amplia cartera de clientes. Joshua se trasladó a Vietnam en 2006 como director de diseño de Lowe & Partners. Sus aspiraciones se centraron en la creación de soluciones creativas a nivel mundial para marcas emergentes, empresas visionarias y causas solidarias en Asia. Después, cofundó el estudio multidisciplinar Rice Creative en 2011. Ha recibido galardones de Red Dot, Pentawards, A'Design, Graphis, Dieline, y el Type Directors Club, así como un premio IDA al diseño gráfico del año por su trabajo para UNICEF Vietnam.

Fish n Rice

| **Design Agency:** Rong | **Design:** Li Sun | **Client:** Fengfan Farm |

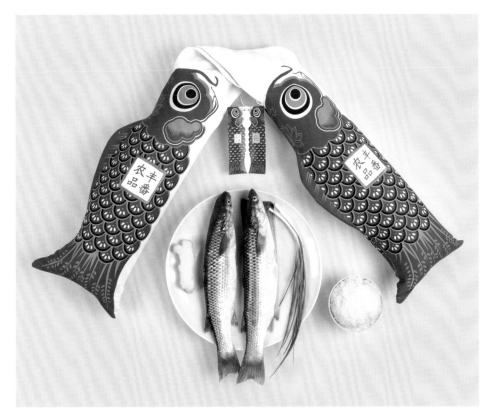

Culture, reuse, and portability are keywords for this 10kg rice packaging design. In Mandarin, "fish (yu)" is a homonym for "abundance (yu)." Traditionally, farmers say "Have fish every year" as a blessing to each other especially during the Spring Festival, which means "Have an abundant year." Now this saying becomes an auspicious wish for everyone.

The one-piece bag design not only emphasizes the shape of the double-fish character, but also balances the weight on each side, so consumers can carry it in a comfortable way.

Sushi Roll Container

The design evokes the lost national and social culture of Japan: the visual design is inspired by a traditional Japanese lantern, while the form encourages the traditional value of sharing. The container also emphasizes on sustainability and portability. The outer shell is reusable, while the paper wrap of the roll is disposable. The outer shell is also collapsible to save storage space.

 +  =

Shiawase Banana

Design Agency: nendo | Collaborator: hnm

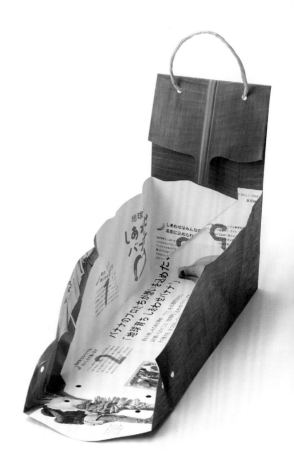

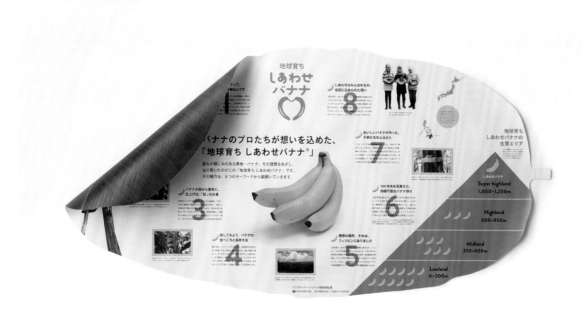

To interpret the environmentally friendly "shiawase banana," nendo decided to steer clear of boxes or packing material, and applied stickers on the surface of the banana peel. The sticker is double layered, with the first layer faithfully reproducing the textural feel and colors of the banana peel. When this layer is peeled off, the second sticker that indicates a message emerges, with an image of the flesh of banana in the background. Nendo created an atmosphere where consumers are compelled to read the story behind the bananas.

Moreover, nendo designed a paper carrier bag made out of one sheet of paper in the shape of a large "banana leaf." The bananas can be easily picked out when the string handle is removed.

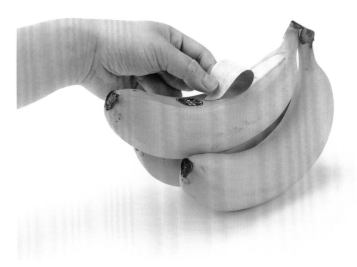

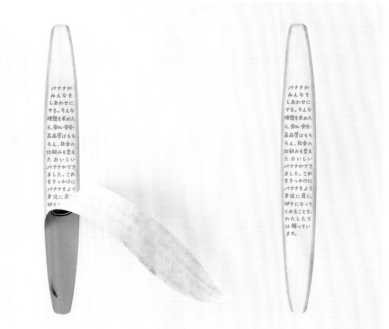

Sweetalk

Sweetalk is the language of temptation for a sophisticated palate! It is a brand of unique popsicles. These succulent iced sweets tickle the senses and unravel a whimsical explosion of refreshing flavors. The brand consists of six different frozen sticks made with fresh and natural ingredients. Each stick is given a lovely name — Lychee Lover, Juicy Melon, Kinky Banana, Sticky Coconut, Oh Honey, and Sinful Berry. The designer creates different packaging system for each flavor which corresponds with the shape of the popsicle.

Pair Champagne

Pair champagne was a school project created by Natasha Frolova and Louise Olofsson. The task was to create a new and innovative package for champagne. Natasha and Louise in the end came up with a package that was portable, recyclable, and foldable.

Ontic Jewelry Packaging

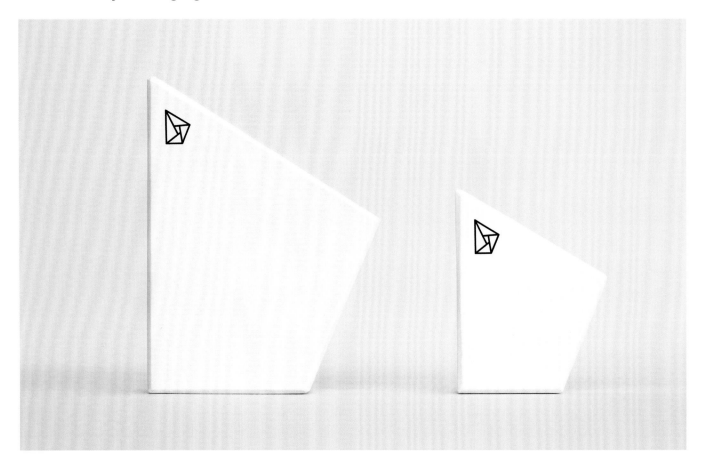

Ontic is aimed at creating personalized jewelry by merging the world of digital visual arts with modern jewelry design. Tasked with the challenge to create an identity for Ontic, Cindy Forster created a simple, bold, and geometric logo constructed with the perfect proportions of golden ratio. To launch Ontic's signature product, a modular women's bracelet featuring geometric and interchangeable artwork, Cindy designed a premium bespoke packaging for the bracelet to reflect the unique shape of the logo. The packaging took on a minimalist and architectural form. Inside the box, an effective canvas was applied to showcase the intricate artwork on display. The designer created boxes in two sizes. One was to contain the bracelet and the smaller one to display the removable artwork.

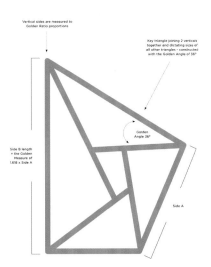

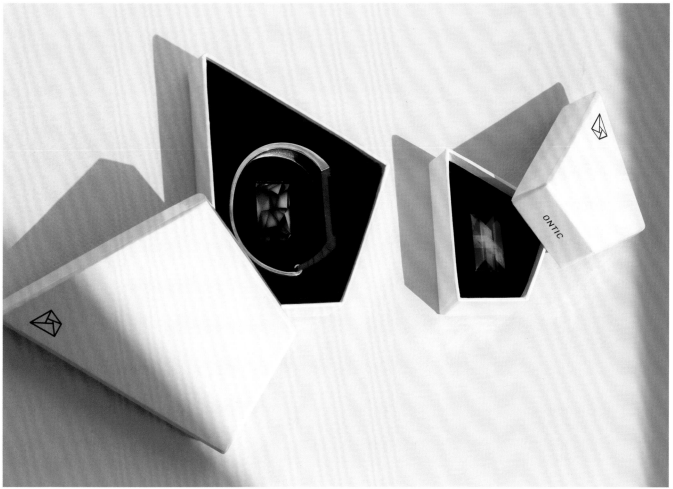

A box of chocolates is designed to look like a set of oil paints from the inside out. Tubes in a box of paints contain a variety of colors, and these chocolates a variety of flavored syrups. The labels indicate each chocolate's flavor and also function as wrappers, keeping one's fingers clean for eating. This is a design that combines the childhood excitement of opening a new box of paints and the thrill of opening a box of chocolates.

Piano Cake Packaging – MARAIS

Design Agency: Latona Marketing Inc.

Japan-based firm Latona Marketing Inc. transforms Patisserie Perle's cake packaging into keyboards. Fifteen cake boxes are lined up neatly to form two octaves. Latona Marketing Inc. cuts costs by focusing on only one box design, and through making use of all six surfaces of the box, they recreate every type of keyboard. In addition to cake boxes, they also create bags which can form larger keyboards. With this design, they are able to create keyboards of any size, from small keyboards to full 88-key grand piano or even larger.

TheWonderfulSocks

Design Agency: Zup Design

This package is made to travel around the world. The peculiar triangle shape and the dimension are wisely conceived for all sort of shipping. The box is assembled from one sheet of black thick paper without using paper-clips or glue. White silk screen printing is applied to highlight the logo.

Møller/Barnekow

Design: Rasmus Erixon, Tobias Möller

Møller/Barnekow is a fictional well-established sandwich bar in central Malmö. For many years they have been offering high quality sandwiches and salads with locally produced ingredients. They now intend to expand their range and offer high quality wraps. The designers' mission is to create a packaging solution that follows the new lifestyle trend "Smart on the Go." A wrap in folded paper easily gets messy and greasy, and it is hard to hold or eat from in a simple way. By contrast, the designers make Møller/Barnekow's package easy to open and carry with. People can simply eat from it on the go.

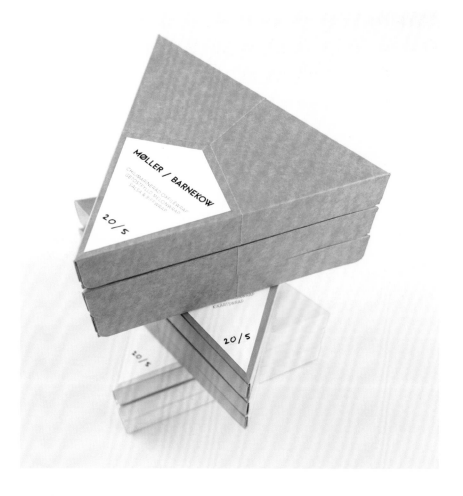

Photography: Rasmus Erixon, Tobias Möller

The project's title was the date of the event, 281214. The source of inspiration was the triangle. A shape perfectly suits the occasion since the event, a wedding-christening, requires the participation of three persons.

This project was more like a combination of invitation and poster. It not only communicated the event's details but allowed people to meet their families by using pleasant infographics and information. The folder-box, also excessive in size, reinforced the concept of the triangle with its pyramidal shape. The use of large typography and industrial materials highlighted its characteristic features.

Cow Country Farm

Design: Spencer Hill

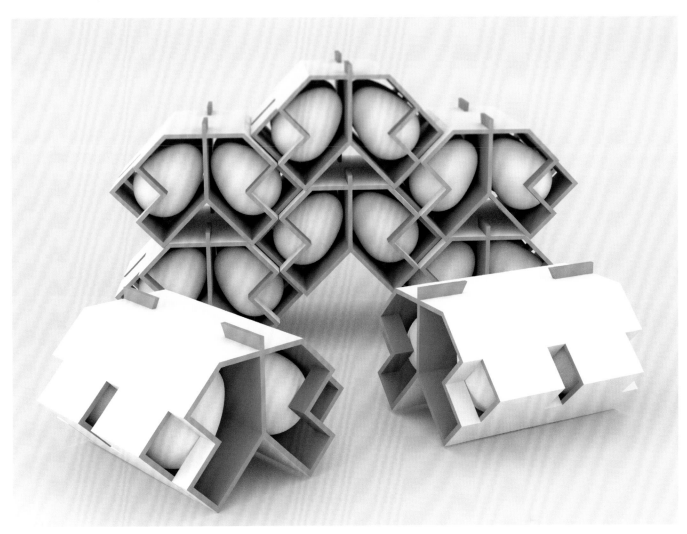

Cow Country Farm is a sustainable packaging solution that turns a single sheet of cardboard into an unconventional egg carton. The design needed to be easily constructed on the farm, while strong enough to transport four duck eggs to a nearby farmers market. The final package uses no adhesive, minimal waste, recycled materials, and hexagonal stacking. To remove an egg, customers simply flip one of the tabs along the outside edge of the packaging.

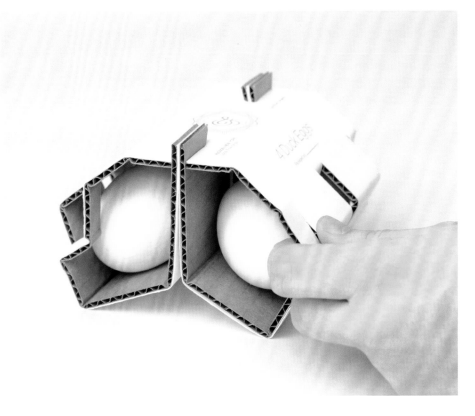

Egg Box

Design & Illustration: Ádám Török | **Photography:** Csaba Szalai

Ádám Török created this package for a packaging design competition in focus groups. The theme of the competition was to design an egg carton for six eggs. He came up with a type of box that was easy to carry, transport, and stack up. It was designed to be recyclable and durable. Buyers could view the eggs from the hollowed-out typefaces.

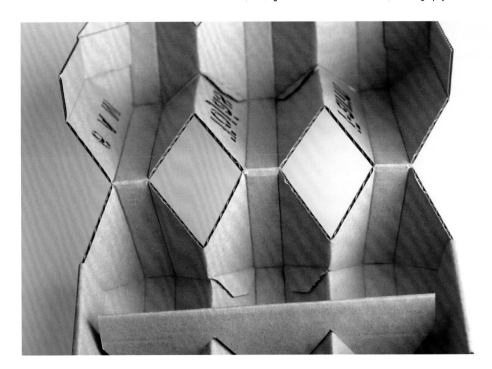

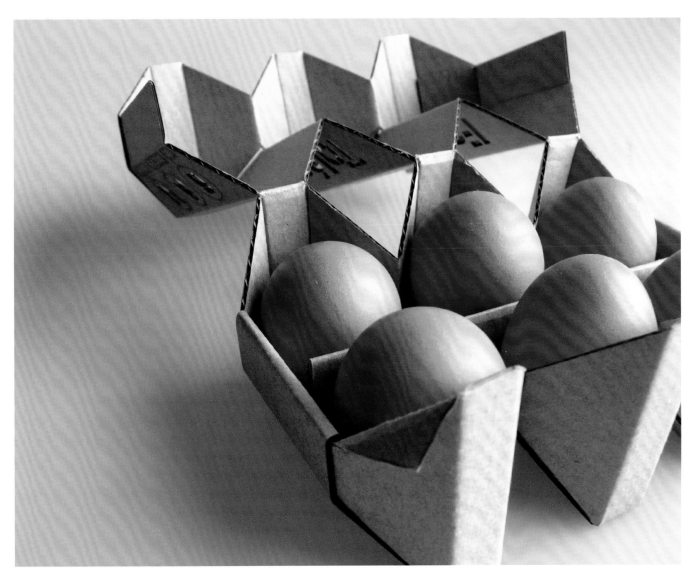

Art Direction & Design: Michalis Kanonis | **Client:** EMBIODIAGNOSTIKI | **Photography:** Michalis Kanonis |

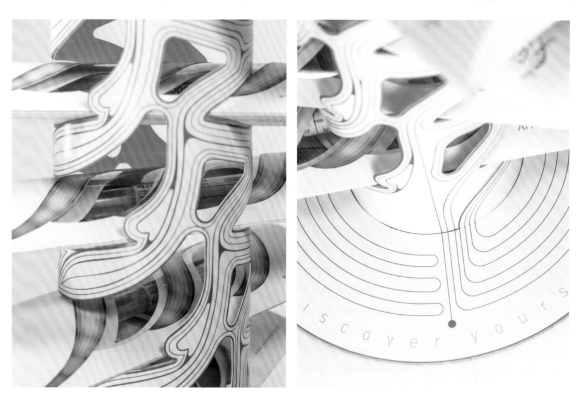

GENOSOPHY (discover yourself) is the name of the genomic predisposing service provided by the Greece-based EMBIODIAGNOSTIKI company.

The three-dimensional design approach of the promotional stand and the single packaging box together refer indirectly to the genetic research through the use of the double helix. A futuristic element is added to the overall layout. The cutout also creates a unique creation, and at the same time promotes the allusiveness of the design with an imminent revelation. The design is developed for the benefits of ergonomics, easy placement, and removal of the packages.

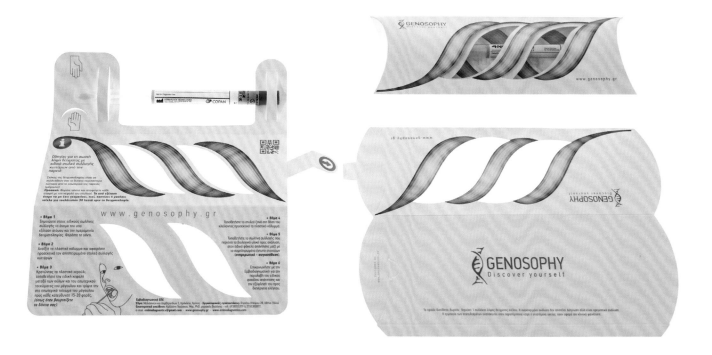

The Pavilia Hill Party Invitation

The Pavilia Hill is a luxury bespoke residence in the heart of Hong Kong and curated by cultural entrepreneur Adrian Cheng, which reflects the Artisanal Movement organized by New World Development. The design pays tribute to nature and artisanship. Its interior and landscape designs are all based on Wabi-Sabi, a Japanese aesthetic and worldview centered on the acceptance of transience and imperfection. Toby Ng designed a bespoke premium invitation for The Pavilia Hill Party, whose theme was "Serenity Above, A Journey of Wabi-sabi." The Invitation was designed as a gift box with a blind deboss abstract brushstroke created by the Pavilia Hill's landscape architect and Zen priest, Shunmyo Masuno.

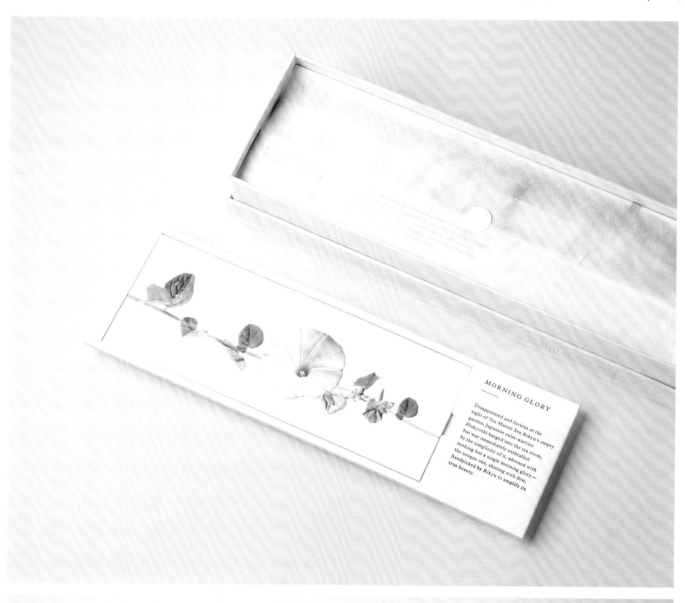

MORNING GLORY

Disappointed and furious at the sight of Tea Master Sen Rikyu's empty garden, Japanese ruler-warrior Hideyoshi barged into the tea room, but was immediately enthralled by the simplicity of it, adorned with nothing but a single morning glory— the unique one, shining with dew, handpicked by Rikyu to amplify its true beauty.

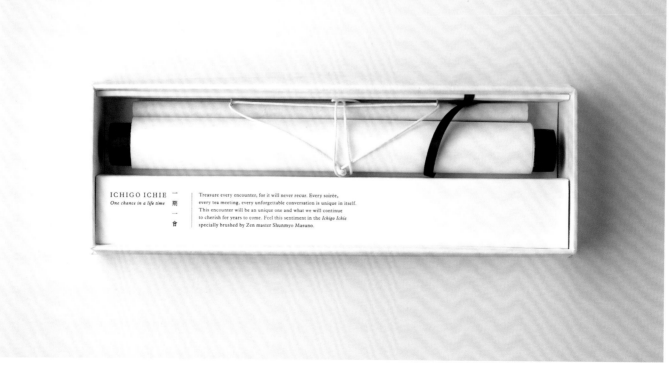

ICHIGO ICHIE   一
One chance in a life time   期

一
合

Treasure every encounter, for it will never recur. Every soirée, every tea meeting, every unforgettable conversation is unique in itself. This encounter will be an unique one and what we will continue to cherish for years to come. Feel this sentiment in the *Ichigo Ichie* specially brushed by Zen master Shunmyo Masuno.

La Casita Blanca

La Casita Blanca is a kind of chocolate
with a bit of mystery. Against the white
packaging, the rich brown chocolates stand
out boldly. The name of the chocolates is
hidden and only reveals when a small tab is
pulled. The name of this project comes from
the La Casita Blanca in Barcelona, which
was founded in the early 1900s and became
a shelter for clandestine love during the
postwar years. Similarly, La Casita Blanca
chocolates look pure and simple on the
outside, but the chocolates themselves are
sinfully delicious.

Create Your Own Moon

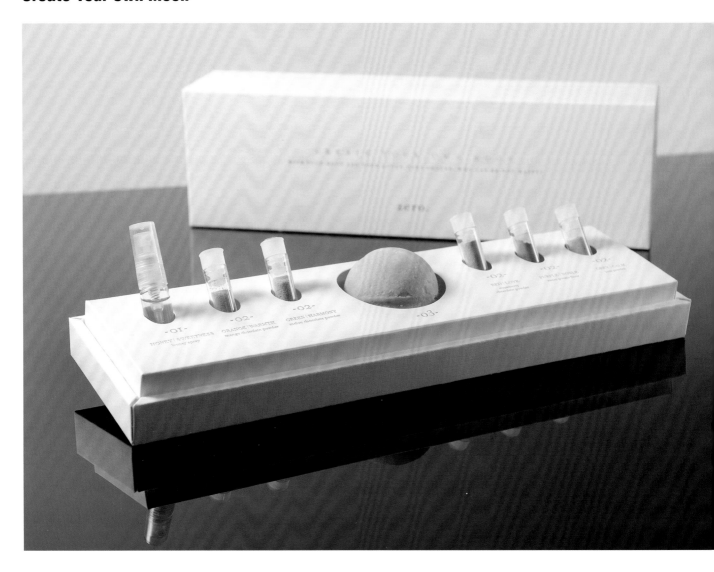

This project builds on the concept of "A Moment of Reunion" during the Mid-Autumn Festival. The users can create a miniature moon with their beloved ones to share the sweetness of reunion implicitly. The colorful powder drizzled on the unique mooncake symbolizes the creativity and vibrancy. Such interactive design adds a new dimension to the traditional activities of this festival.

Citrus Moon

Art Direction & Design: Tsan Yu Yin

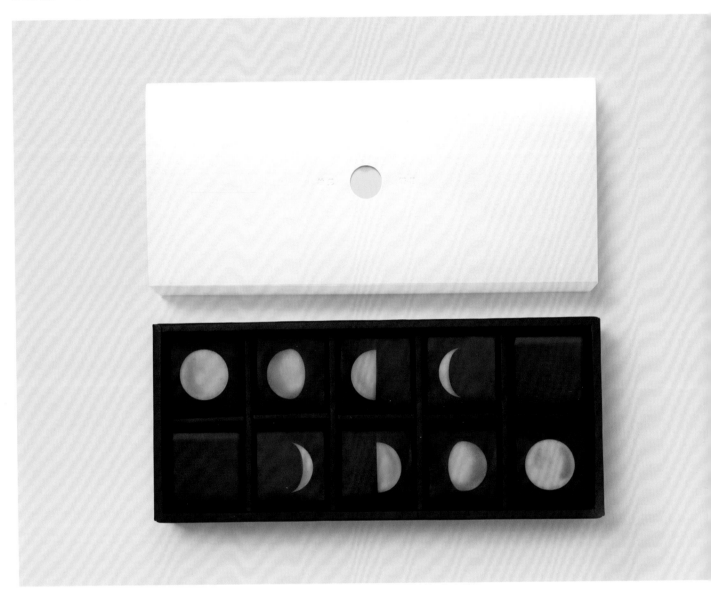

Citrus Moon is a gift box for the Mid-Autumn Festival. The packaging of Citrus Moon explores both the context and colors of the oriental heritage. On the exterior sleeve the gradient symbol represents the mid-autumn full moon as well as the shape of citrus. The progression of lunar phases will be revealed through the die-cut circle when the sleeve is being pulled out. Packages of the ten mooncakes represent respectively different lunar phases, and they are arranged in the order of a whole lunar cycle.

Let's Make A Wish

Design Agency: invisible | **Design:** Lam Cheung

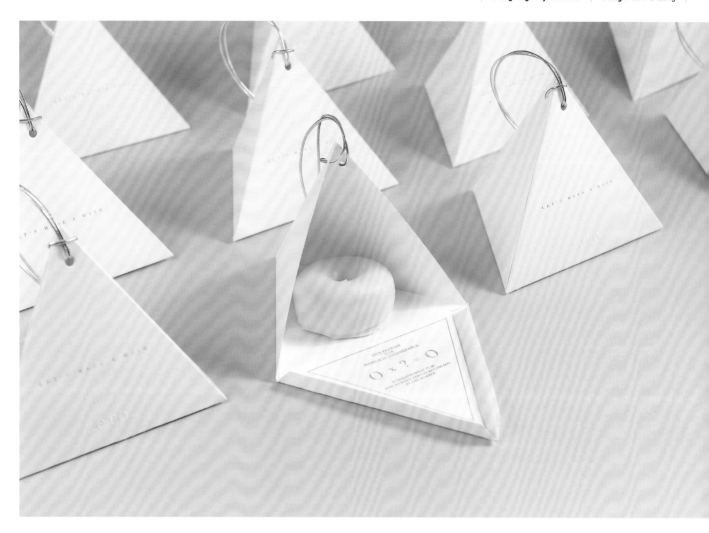

Christmas might not be an eco-friendly holiday. From seasonal decoration to food sector, lots of trash is produced during the Christmas, especially package trash of different items. The studio intends to contribute their effort to solving this problem. They design an all-white gift box on which people can make a wish and turn it into a decoration on the Christmas trees. Now let's make a wish!

Titleist Lunar

Design: Jens Marklund

This is a conceptual packaging in honor of Apollo 14 and Astronaut Alan Shepherd's attempt at playing golf on the moon. The solid base is made of steel and is given four branches at the middle where the golf rolls are easily hung. Each roll holds three golf balls firmly without any adhesive. Customers can open them up with a simple twist. The purpose of the structure is to create an object that users would want to keep and reuse.

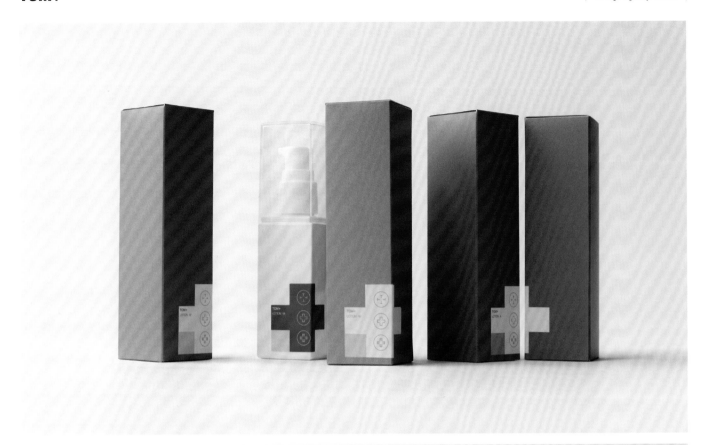

TCM+ is a branding concept for a new line of skincare products to be launched in Asia. Formulas of the brand can be customized. Instead of being supplied as premixed products, the idea of TCM+ is for the end-user to freely customize the mix of ingredients and relative quantities according to their body's needs, the time of the day, or season. The name of the brand, "TCM+," is a reference to the process of "putting together" various cosmetics with traditional Chinese medicinal properties, while the logo appears as a "+" through placing in sequence the letters "T," "C," and "M." Additionally, nendo came up with a package design "+" that appears when lining up the various products with different properties on a shop or cabinet shelf.

| Collaborator: nsz | Client: TCM+ | Photography: Akihiro Yoshida |

Good Hair Day Pasta

Nikita Konkin uses the strands and shapes of pasta to create an interesting series of pasta packaging that captures the attention of customers. The design emphasizes the high-quality and naturalness of pasta. The whole series consists of three packages in different styles with an aim to bring good mood as well as good taste for shoppers.

KADOKUWA Kanroni

| **Design Agency:** Ono and Associates Inc. | **Design:** Takeshi Ono, Ayako Ono |

Ono and Associates Inc. is commissioned by KADOKUWA Inc. to create a package for their product Kanroni. Kanroni, traditional Japanese food, is a type of fish boiled with soy sauce, sugar, sweet rice wine, and sake. Ono and Associates Inc. proposed a vertical package with simple shape and structure. Traditionally the package for Kanroni is to prevent the secret broth from leaking. As KADOKUWA Kanroni was particular about the texture and technology, the designers intended to equally present such texture on the package. Thus, they adopted the idea of letting broth sink into traditional Japanese paper, and applied blind stamping and gold leaf stamping to highlight the product's premium quality.

Zico Coconut Water Rebrand

Design: Ragini Sahai

This is an experimental rebrand of coconut water packaging for ZICO Beverages, LLC. Ragini Sahai redesigns the original bottle form, logotype, graphics, and multipack carrier, and comes up with a brand new package using paper and rope bag. The new package resembles the fresh, green coconuts from which the beverage is sourced.

Puritea Multifunctional Packaging

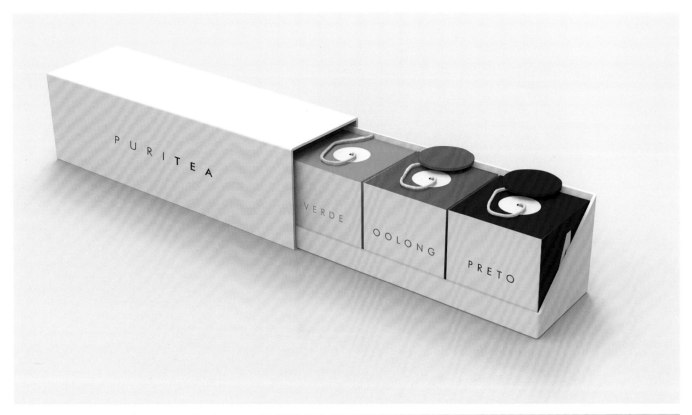

The project is designed to preserve the tea history and culture of the tea ceremony, as it contains four types of teas that are considered to be "pure," namely white tea, green tea, oolong tea, and black tea. All the four teas are produced from the Camelia sinenses leaf. Vinicius Hideki creates the identity and packaging for Puritea based on the necessity of the product. He selects colors according to the tea strength, from the light green to the dark blue. The individual package for each tea is constructed based on origami techniques, the art of folding paper. The inner side of the cube is coated with a layer of wax so the steam will not damage the paper.

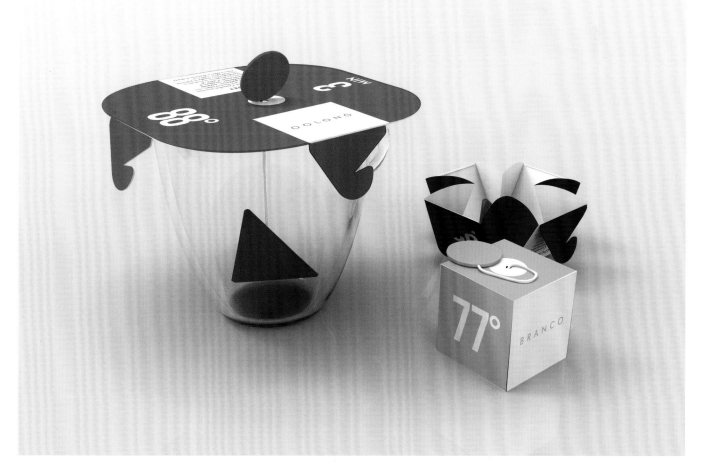

Kiwi to Go

| Art Direction & Illustration: Irene Acosta | Label Design: Irene Acosta |

Kiwi to Go is a school project designed by Irene Acosta and Ágnes Gymrei. Plastic bags account for a major part of daily waste in landfills and oceans. The challenge is to create a design that breaks established packaging rules and helps the environment at the same time. The designers create a natural and organic package that gives the fruit a fun high-end appeal, which is suitable for exclusive organic grocers, markets, and gourmet boutiques. This package is 100% reusable, recyclable, and biodegradable.

Flashtones

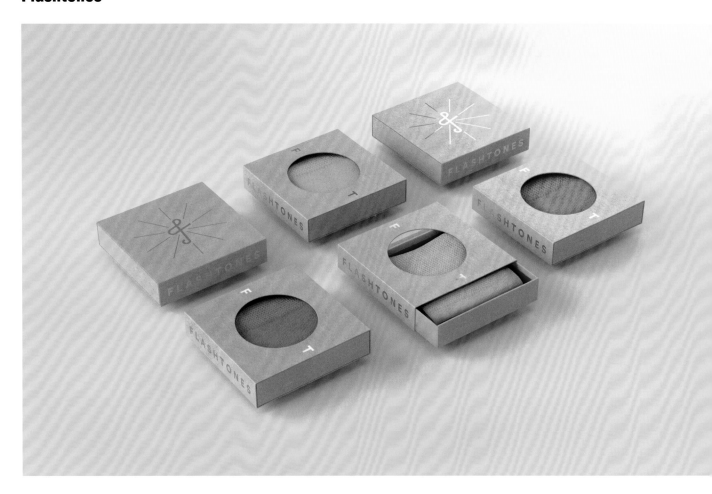

Flashtones is a fresh brand based in the Czech Republic producing colorful socks sold as single pieces. The brief was to create a playful visual identity along with a packaging design for a single sock pack. The main goal was to simplify the in-store customer experience focused on a quick and easy choice of desired colors, with production cost as low as possible. Lilkudley was also told to design a mystery packaging, containing fourteen randomly colored socks. The idea behind the mystery packaging comes from the well known paper towel box, but instead of pulling out paper towels, consumers pull out random color socks. With the mystery packaging, consumers will never again have to spend their morning deciding which colors to wear today!

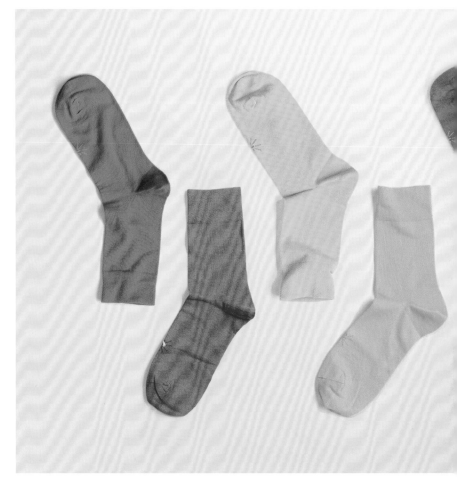

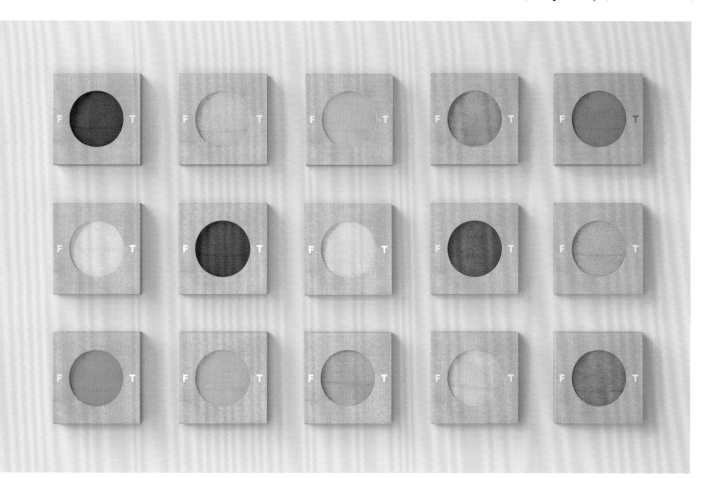

Design Agency: mousegraphics | **Creative Direction:** Greg Tsaknakis

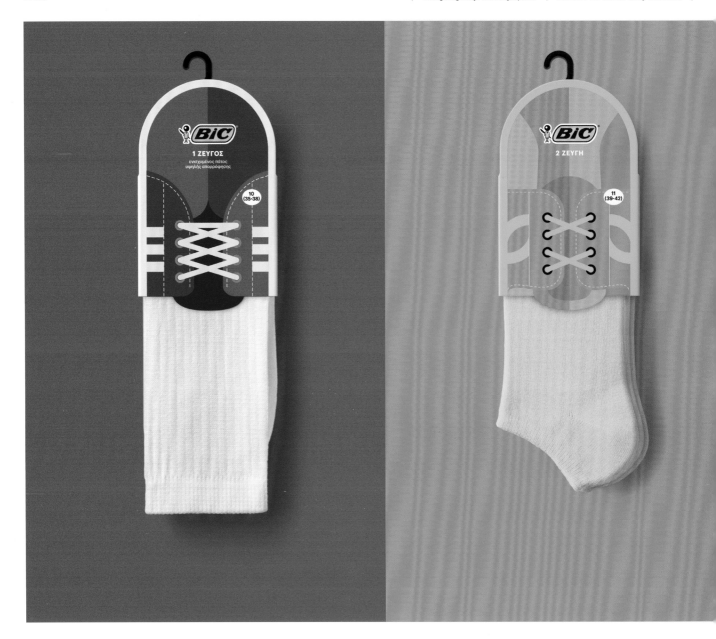

Commissioned by BIC, a well known company with a strong brand, mousegraphics proposesd an innovative packaging idea for the brand's various kinds of socks. Designers intended to refresh the sock's image through packaging and also unify several products under one smart idea with functional variations. Commonly, socks are dressing the feet and shoes are completing the process. But mousegraphics took a mental step ahead and they dressed socks in shoes thus creating a playful visual. Each sock packaging is the image of a type of shoe paired with the appropriate sock, ranging from women's, sports', men's, to casual ones. BIC's products are differentiated, but a larger, and easier to "read" category, namely shoes, is now providing the unifying principle.

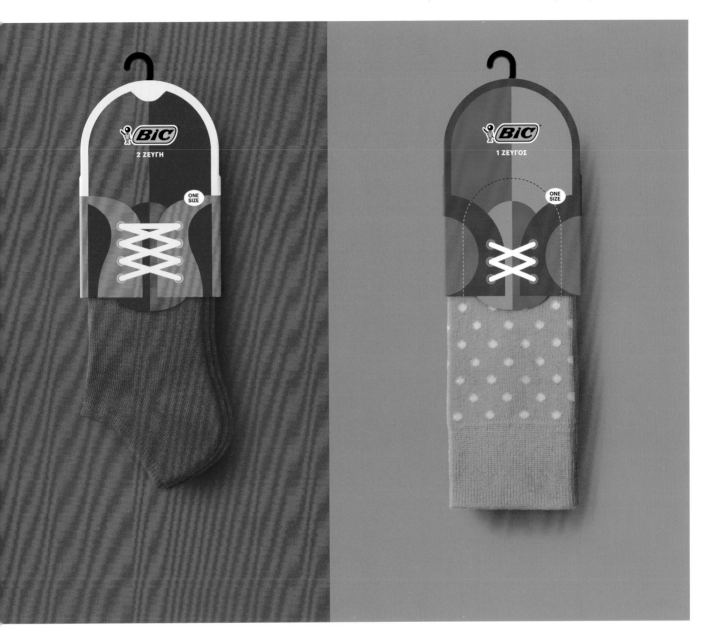

Chris + Jessie

This is a school project which is based on the two founders, Chris and Jessie. Hyela Lee takes inspiration from Chris' love for stripes, Jessie's love for dots, and their collective love for socks, and leads to the birth of Chris + Jessie. Chris and Jessie hope their extraordinary socks bring life and fun to the mundane reality.

To respond to the founders' ideal, Hyela used mundane objects such as a window, a banana, and a cup of tea with a black half-tone to create a vivid and fun package for each sock. The logo "C+J" is a custom letters from Avenir and Proxima Nova fonts and the designer uses INT to add spot-gloss feeling.

SOC TOKYO

Design: Keiko Akatsuka

SOC is a Japanese brand of socks whose slogan is "An inspiration born from your foot." The package is designed to be able to contain socks of all sizes and display the different styles. The designer draws inspiration from the art of traditional Japanese folding and creates a package which users can open it easily without destroying the packaging. The result is compact with the minimum use of paper. It is reusable, inexpensive, yet attractive, which is much like the value of SOC itself.

Flora

Design Agency: Iglöo Creativo | **Art Direction & Design:** Paola Coiduras

To celebrate the 30th anniversary of the International Theatre & Dance Fair of Huesca, Iglöo Creativo created Flora, a packaging design for the commemorative T-shirts. After designing the image of the event in which a tortoise called Flora was the protagonist, Iglöo decided to create a special packaging by using cardboard, plastic, and vinyl. They created a unique package with the form of tortoise to echo the image of Flora in which the shell was used as container. Through the transparent shell, customers could see the logo of this special edition.

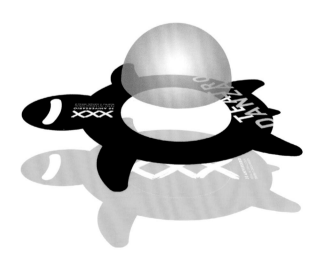

Pandle Repackaging

Pandle is a portable, rubber handle infused with nanosilver, an antibacterial material. It is perfect for holding onto surfaces in public places. The challenges are to explain the product's use and benefits, to show how colorful the product is, and to present the product's touch and feel. Hani Douaji creates a new packaging that is inspired by the use of the product. On the foreground the die-cut has a hand shape which is inserted physically into the product, while on the background there is a simple illustration of a handle. The packaging is designed to look like the die-cut hand gripping the handle illustrated in the background.

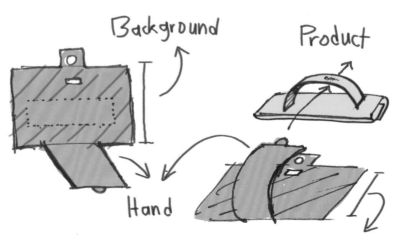

The link between
the front & the back side

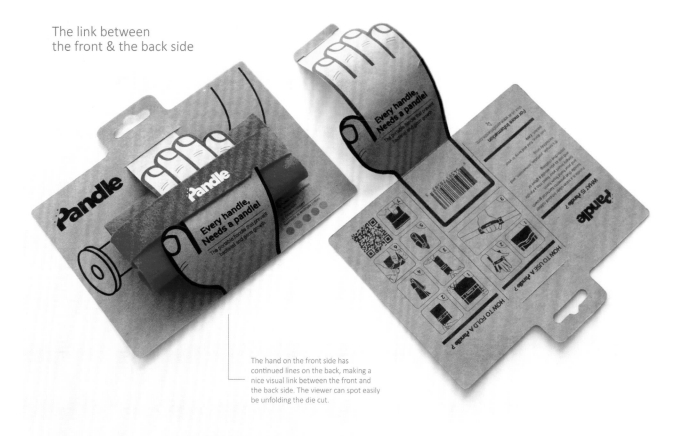

The hand on the front side has
continued lines on the back, making a
nice visual link between the front and
the back side. The viewer can spot easily
be unfolding the die cut.

Exotic Snacks

Design: Rasmus Erixon

Exotic Snacks is a natural candy assortment that gradually becomes a known brand in Swedish convenience stores. Today's society is becoming more health conscious and careful about what they and their children eat. Therefore, Exotic Snacks now wants to develop a concept that makes more children choose a healthier candy. The task was to develop a range of Exotic Snacks natural sweets for children. Rasmus Erixon designed three illustrations of monkeys using patterns to make the candy's package look more playful and attractive. The colors reflected the flavors of the candy.

SKINS Shoe Box

SKINS Shoe Box explores how the traditional shoebox can be repurposed into a more meaningful experience with multiple uses. The project addresses the need to design a reusable and multifunctional item that promotes sustainability and reduction of waste. The all-in-one design functions as multiple tools: portable box, stackable shoe shelf, and hanging wall organizer. The shoebox itself acts as a storage box which can be stacked together with multiple boxes to create a shoe shelf. When users fold out the portable box, the box reveals itself as a narrow wall organizer made out of strong paperboard. With a fabric handle on the box, customers can hang it behind a door or in the closet.

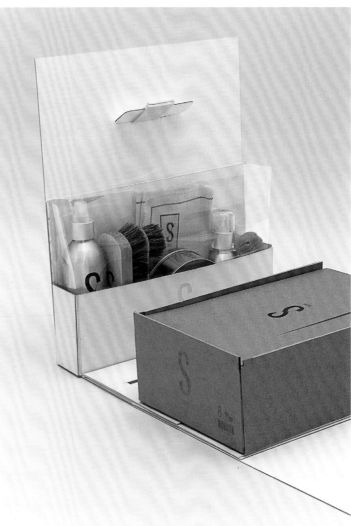

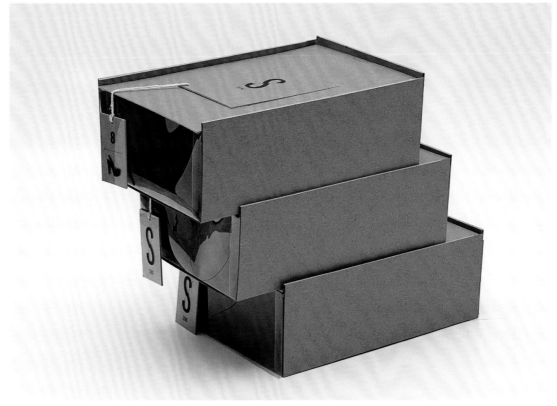

Design: Velimir Andrejevic, Sofija Gavrilovic

Illustrations: Velimir Andrejevic, Sofija Gavrilovic

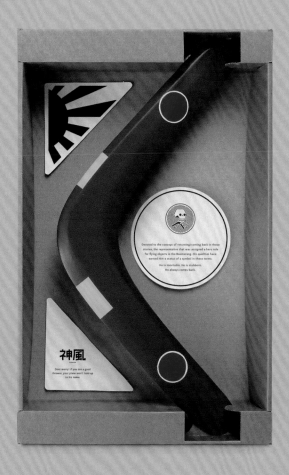

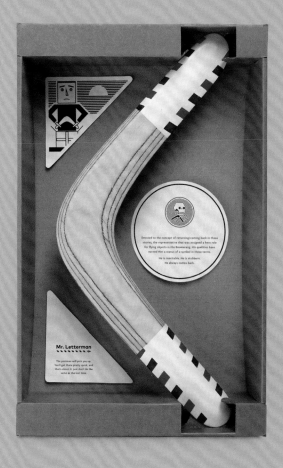

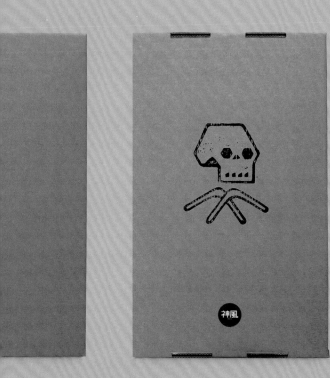

The Returning of the Boomerang is a self-initiated project created by Velimir Andrejevic and Sofija Gavrilovic. It consists of short stories and flying objects — the Boomerang. The packaging was conceived as both a storage box and a wall frame when the boomerang was not in use. It could be easily hung on a wall with the boomerang suspended in it as an art piece.

MATCHPOINT

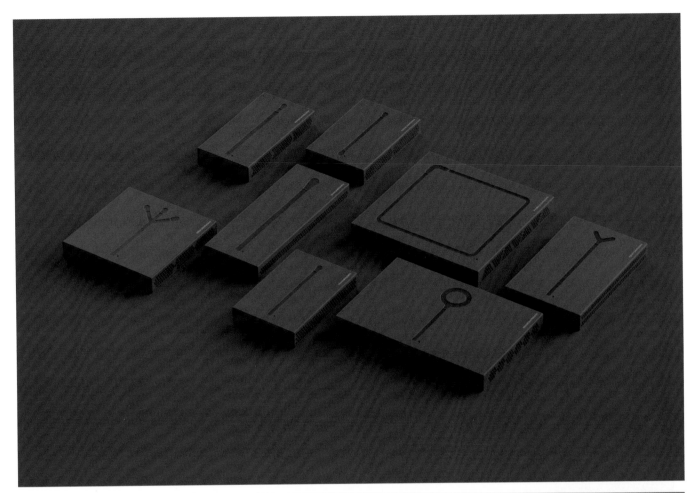

There are many things that are or being forgotten because of some different reasons. Some become less necessary and less used, simply because they do not perform better than the new alternatives. Matches are one of them. People still use them, but not as much as before. Nevertheless, designers believe that matches still worth existing if people recall the magical moment of igniting a match by striking it with a proper snap.

To light up the underestimated preciousness of matches, designers explored different cases of lighting up fire with matches, such as double-sided matches, torch matches, matches for lighting a fireplace or a cigarette and so on.

Design Agency: twelvemonthly   **Design:** Dongkyu DK Lee, Junghoon Lee

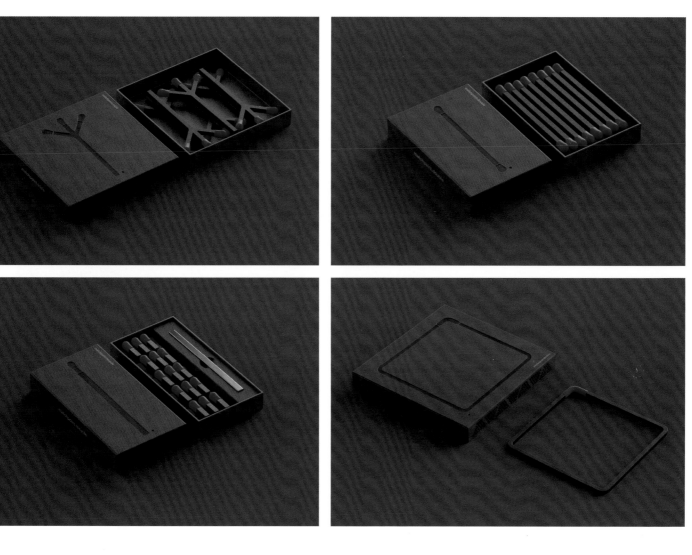

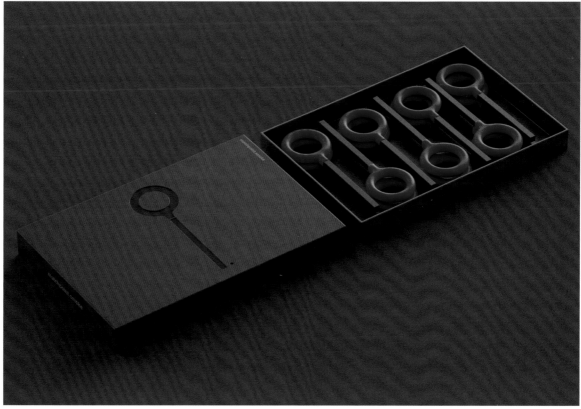

Rope Pack

Design: Isabella de Felice | Illustration: Annachiara Barindelli | Client: IF BAGS

Rope Pack is a self promotional project for IF BAGS. Ropes in various colors are one of the most important products of this brand. Customers can choose the desired colors and change it by themselves. The rope and the illustration on the package go with one another as the rope inside becomes the beard of the male character.

One iOS Lightning Charger

Design Agency: MADE | **Design:** Veronika Levitskaya

One is a small boost battery developed especially for urban citizens. One will easily live throughout the day, keeping the phone charged at the desired 70%. MADE studio designed the logo, identity, and packaging for this tiny power bank. The packaging is very compact and convenient, and so is the gadget itself. It can be opened in a fast and easy manner, and attracts attention to the original bank (case) form. It resembles the letter "O," which resonates with the key element of the brand identity and the oval shape of the bank. It is not only a handy gadget but a perfect gift with functional and fashionable packaging.

Salmon Oil

Design Agency: mousegraphics | Creative Direction: Greg Tsaknakis | Art Direction: Kostas Vlachaki

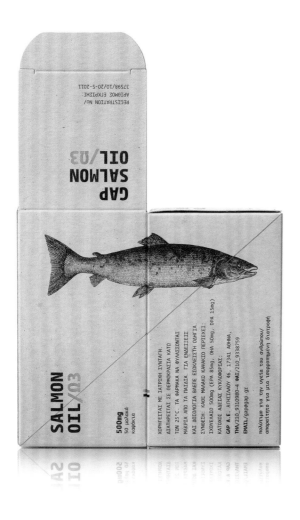

GAP's Salmon Oil is a valuable nutrition additive and it is looking for a distinguishable package. This product is dispensed by prescription and addresses people's need of vitamin based nutrition additives. Mousegraphics opted for a design which would be a direct and rather sober reference to the very natural source of the product, a salmon fish. They chose the iconographical approach of a fish encyclopedia of the 18th or 19th century, coupled with the simple, clear typography. The devise of the orange diagonal line would connect the packages when displayed.

Springs' Smokery

| **Design Agency:** Distil Studio | **Design & Illustration:** Neil Hedger, Fiona Curran |

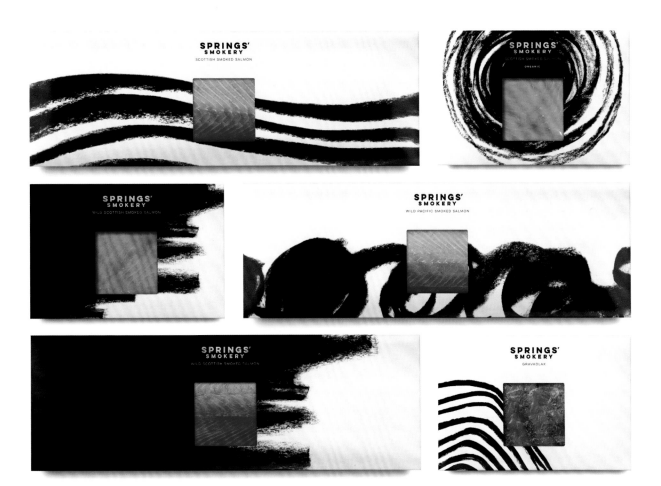

The rebrand for Springs' Smokery is drawn quite literally from the heart of Springs' craft. Distil Studio takes the charcoal embers of oak logs from their 50-year-old smoking kilns, and uses them to conjure a series of emotive marks and textures. These include charcoal expressions of the surrounding South Downs, Scottish rivers, and the waves of the Pacific. The supporting copy focuses on Springs' proud story and their ethos of honest food with no additives. The result is to present the ancient tradition of smoking using a contemporary design language, which is bold, dramatic, and rich in handcrafted values.

Happy Super 2016

Zup Design is asked by CTS Grafica to create Christmas kits for gift wrapping. The designers come up with wine and oil bottle packaging printed with fluorescent colors and Pantone metallic, along with greeting cards and double-sided wrapping paper. Happy holidays to everyone!

| **Design & Illustration:** Andrea Medri, Riccardo Pierassa | **Client:** CTS Grafica |

Design Papers 2016

| Design Agency: Metaklinika | Project Management: Nenad Vulović |

Design Papers is a paper catalogue created for Europapier, a company with over 40 years of tradition, distributing paper in more than 13 European countries.

The idea was to add a decorative dimension to the usual functionality of a catalogue, with attention to both its visual and tactile aspects. The catalogue contained 240 types of paper. It was designed as a box divided into two colors and two groups of pictograms with a rhombus in the middle. The title "Design Papers," positioned in the rhombus, was intersected by two areas. The rose area contained the first group of pictograms "Look, touch and feel," emphasizing the interactive character of the catalogue. The other group of pictograms had a gold foil finishing on a turquoise background, and the words "Carefully created collection" emphasized the excellence of the catalogue and its outstanding quality.

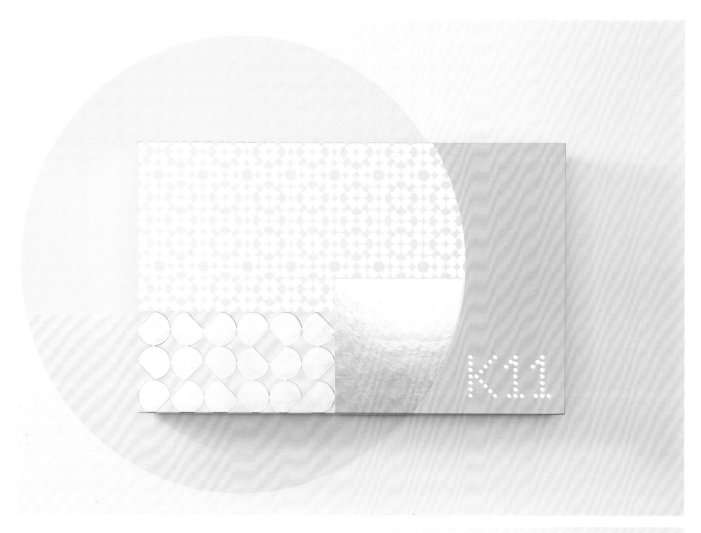

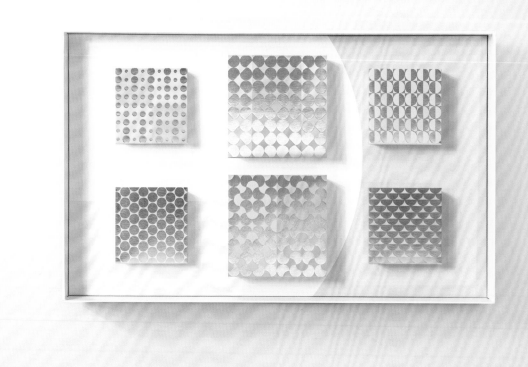

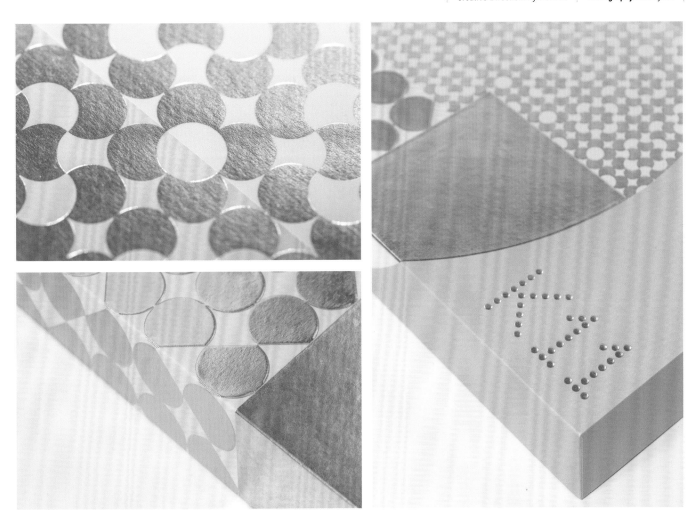

To celebrate the Mid-Autumn festival, Not Available Design develops the packaging for K11 Mooncake. These tasty cakes are made of bean paste and crisp crust. Since they are super sweet, people often enjoy them with a bit of tea. The designers use abstract graphics and contemporary colors to represent the lunar cycle on the boxes and bags. The lovely pastel colors can be interpreted as the lunar cycle and colors often seen as a halo around the moon in the sky, or reflected in the water. Shining silver foil is used to add a celebratory element and give the package a sharp appearance. The patterns are soothing and relaxing which also represent the changing shape of the moon in the night sky.

Bonbonbon Wooden Box

Design: Zsofi Ujhelyi, Diana Ghyczy | **Bonbons & Photography:** Dorka Meleg

The Bonbonbon, hand-made coconut balls, is a gastro project created by three friends. Walnut, almond, coconut oil, and seasonal raw materials are used. Bonbonbon needed a new packaging system in which they could be stored and served in coffee shops as well. The result of the designers' joint forces became a hand crafted bonbon box made from cherry wood, while the paper take-away packaging was coated with greaseproof foil from the inside. Each bonbon has a unique package, design, and taste, making these sweets a special experience.

Twist Packaging

| **Design Agency:** Communal Creative | **Structural Design:** FORMA | **Printing:** Design Packaging Inc. |

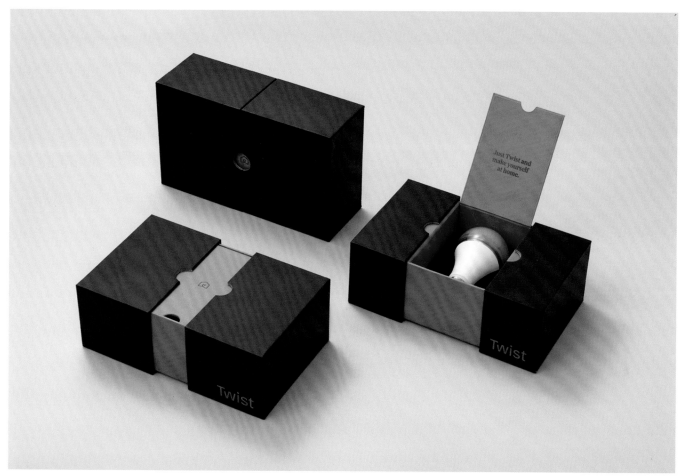

Twist is a smart LED light bulb that combines adjustable lighting with high-quality audio, via a built-in AirPlay speaker. The seamless technology removes the annoyance of hubs, wises, and installation, providing a simple solution for a smarter home.

With the elegance in product, Communal Creative elevates the delivery experience by creating a packaging system that adds polish and delight. The unveiling journey is mapped by the rich brand colors, playful illustrations, and unique product display with a footprint that doubles in size once opened. The material is the key to final execution — each box is finished in a soft touch paper, creating a great hand-feel.

Desig de Sant Joan

Desig de Sant Joan, meaning Saint John's wish, is a premium hand crafted line of chocolates seeking to capture the most important night of the year in Barcelona, the Saint John's Eve. The designers create this project for their Master's degree in packaging design at ELISAVA university.

From the chocolate and packaging, to the brand and graphic identity, each element is designed and inspired in the visual components of this magic night, where fire is showcased as the main protagonist in the form of firecrackers, fireworks, and bonfires.

The exterior design reflects the strength and explosiveness of fire through the pattern generated from the laser cut cardboard; whilst the interior packaging wraps a differentiating factor, where the designers create a surprise element contrasting the form of the chocolate and its vibrant colors with the minimalist packaging design.

Design: Claudia Lepesqueur, Maria Vidal, Maria Elipe

Eden & Bridge

The brand name "Eden & Bridge" is derived from the famous paradisiacal garden in which Adam and Eve lived, which also echoes the location of the brand owner — Edenbridge, Kent. It is often said that the Garden of Eden is marked by four beautiful rivers, adjoining together, enabling the vast virility and fertility of this area with unsurpassed natural beauty.

The brief was to conceptualize and create this luxurious, lavish, and seemingly decadent pie brand. Innovative packaging, which encouraged a gratifying experience to the consumers, was of paramount importance in promoting the true extravagance of this new brand.

Cocoa Colony

| Design Agency: Bravo | Creative Direction: Edwin Tan |

| **Design:** Jasmine Lee | **Project Management:** Janice Teo, Jacintha Yap |

Cocoa Colony is a chocolate brand that tells the story of two brothers who discovered the many benefits of cocoa beans when they found themselves stranded in Ecuador after their ship capsized during the Colonial Period. With its healing properties and delicious aroma, the precious cocoa beans became known as the "Amazonian Gold" when the brothers brought them back to their homeland.

Bravo felt the need to retell the story in elaborate details through meticulous typography and material choices during the creation of the brand. Gold was embellished with great intent to emphasize the affection that the designers have for the product.

Cacao Barry – Conduru

Design Agency: Zoo Studio | **Design:** Xevi Castells, Gerard Calm

Cacao Barry's Conduru is an exclusive chocolate that comes from a small cocoa plantation in Brazil. With this cocoa, 100 chocolate cylinders of 1 kg have been produced in order to get a limited series of 100 packages. The first of them, and always the most special, is delivered to Pierre Hermé, who has been named the world's best pastry chef at the 2016 World's 50 Best Restaurants ceremony in New York.

The concept and design of the package come from the idea of taking advantage of the main material that farmers use to store the cocoa, the wood. Thus, Zoo Studio searched different kinds of wood with different thicknesses for the packaging. Some are natural woods, and some other with natural dying.

RING BOX

Design: Leonie Werts

RING BOX is a handmade beautiful present box for jewelry. It has the look of a gift which will attract the customers to unpack it. In this present box there is a valuable ring. The RING BOX is made of two pieces. The two pieces fit perfectly together. The gap inside one of the pieces is especially made for storing the ring. Together the two pieces serve as a gift. It is totally made of white painted MDF and Oak wood.

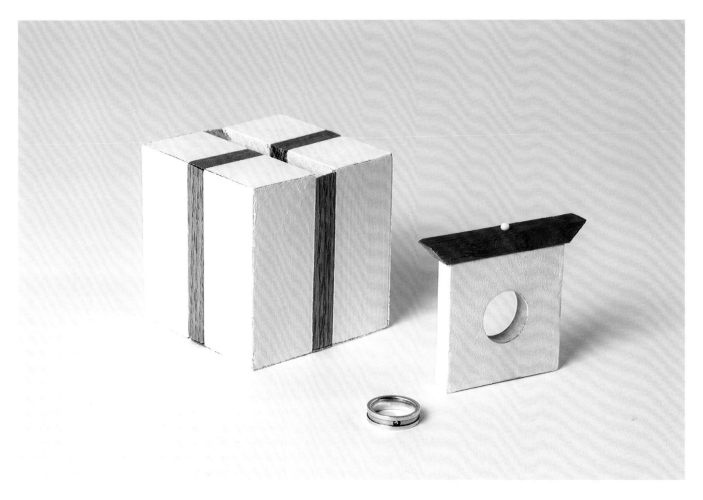

Echo Earbuds Container

Design: Brandon Stacey

This is an industrial packaging project that Brandon Stacey made in his studio class. Echo earbuds was intended to deal with tangle. A magnet was utilized to hold the halves together. A laser cutter was implemented to burn the wood surface with the branding. The finished product is a stylish wood wheel that keeps the tangle free.

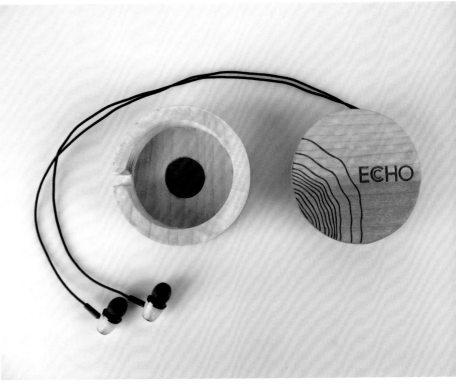

WDSTCK Barber Blade

Design Agency: Tough Slate Design | **Design:** Dima Tsapko, Roman Davydyuk |

The identity is based on the motives of topographic map of the specific region, where WOODSTOCK Sawmill's wood supplies are coming. The overall design of the blade is made corresponding the idea of WOODSTOCK Sawmill's identity. The box design resembles the design of striped barber shop light.

Bzzz Premium

A local honey manufacturer sets the assignment of developing a name, logo, and packaging for their quality natural honey. This time Backbone Branding was given a new task — to wrap the limited edition of honey jars as if they are a business gift.

Designed in a notion of biomimicry, the wooden hive left no room for alternative concepts and made the way to the technical execution. What kind of container can enclose honey better than the hive itself? As for the brand name and the logo, they were developed to resemble the natural buzz and waggle of the bees.

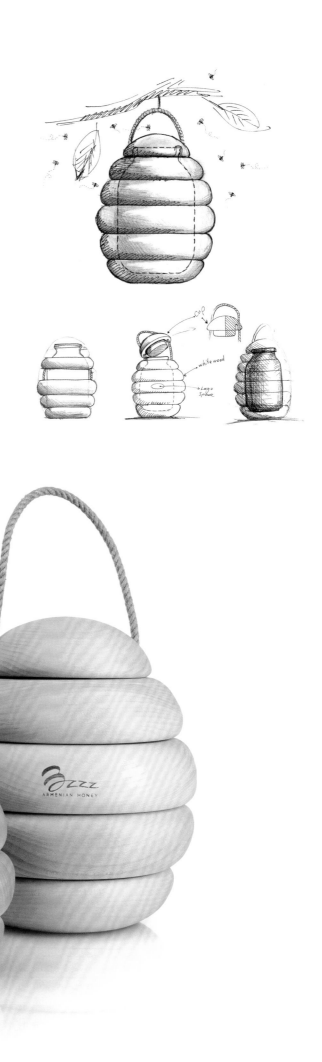

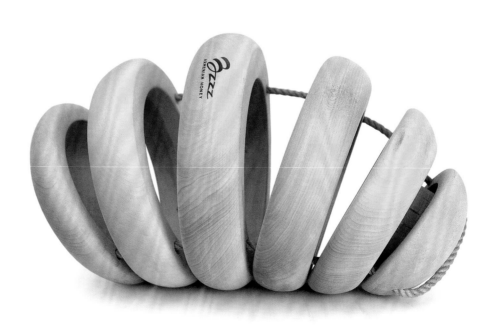

Packaging Stone Pine Liqueur Zirbn

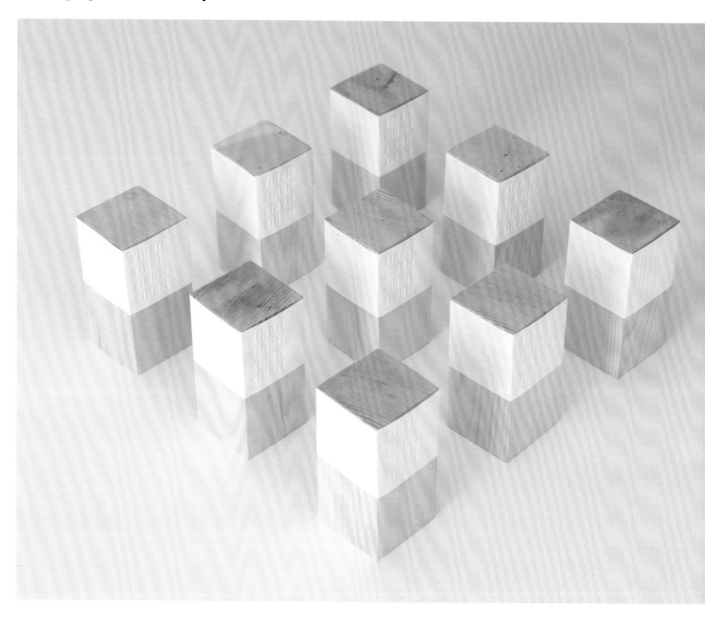

Hermann Knogler's high-quality liqueurs are all handmade. The main focus on the limited edition "Zirbn" is reduction. The liqueur is bottled in a container made of stone pine wood. The label contains only the most essential information and the text is blind embossed. The design emphasizes the senses — the smell, taste, and feel of stone pine.

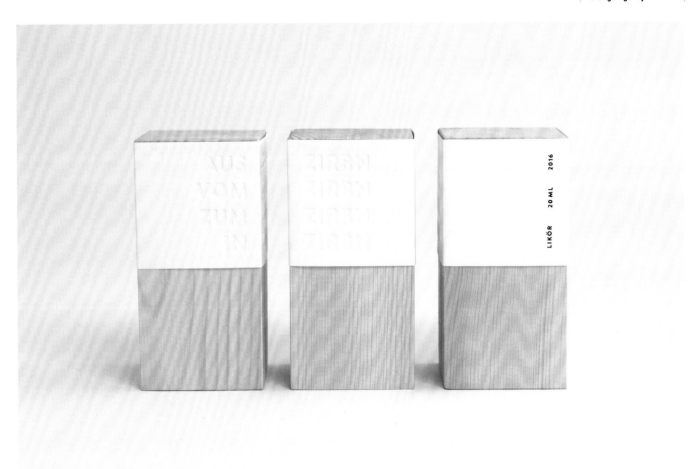

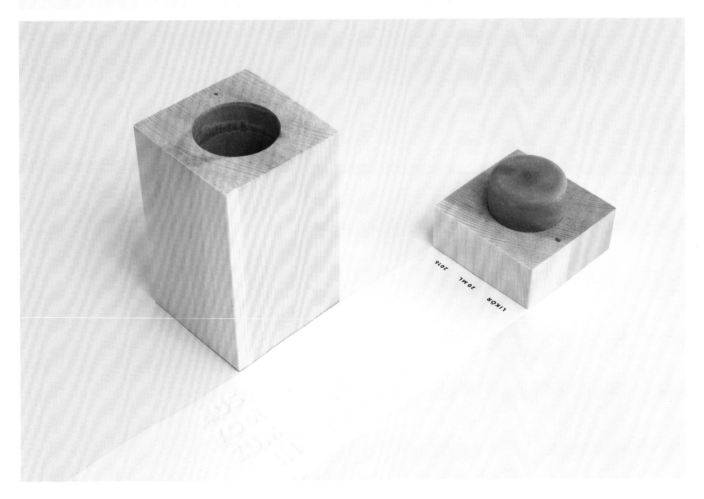

Holcim Agrocal

Design Agency: Studio Sonda | **Creative Direction:** Jelena Fiskus, Sean Poropat

Agrocal powder is a natural and eco-friendly source of Ca and Mg that increases soil fertility. The solution aims to pack the product in ecological, easily degradable paper that uses only one color in print. It is set in a wooden box that is not supposed to be disposed off, but repurposed and reused by gardeners. Different motivational sayings are carved on the box in order to encourage gardeners to a kind of collecting. The principle of functionality with the retail space is the key point — the packaging is designed so that boxes can be stacked one upon another, becoming an independent entity without having to struggle for a place on the shelves in the usually crowded agriculture supply stores.

Koronaki Virgin Olive oil

Design Agency: Artichoke Design | **Design:** Thomas Kiourtsis | **Client:** KORONAKI

KORONAKI is a small Greek company that produces excellent quality virgin olive oil and high-value traditional Greek products. Artichoke Design is asked to design the corporate identity and product packaging for the brand. The packaging is based on two basic materials, wood, and stone, while stone is attributed with ceramic material. The designer combined these two materials into a modern and simple shape, which reflects the uniqueness and high specification of the product.

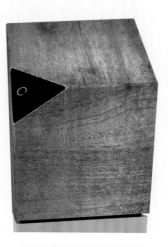

Bevel Cup

Design Agency: NANOIN Design

This is the packaging design of Bevel Cup. This design is made via pulp casting, which is low-cost and convenient for recycling and degradation. Both its internal and external modeling are made by a set of mold respectively, and then both models are bonded as a whole. The external modeling aims at delivering modeling features of Bevel Cup, and the internal modeling is closely embedded into cup body, which can effectively reduce the failure rate of ceramic cup in transportation process. The contrast between internal and external modeling is also a visual language which the design pursues for.

Cacao Barry – Tocantins

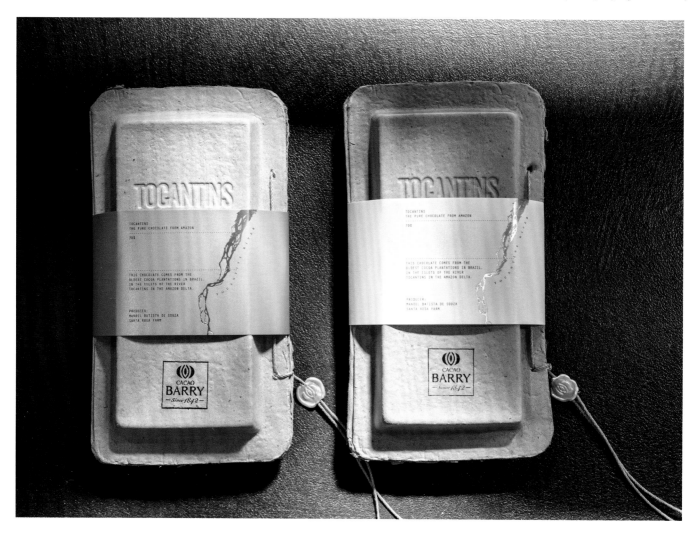

Tocantins is a chocolate that comes from small cocoa plantations from the Amazon delta. The production is limited and reserved exclusively to the "World's 50 Best Restaurants." The packaging for the presentation is made with paper paste, tied with fine rope, and sealed with sealing wax, evoking a remote origin with a band that provides exclusivity. The band is hand signed by Jordi Roca from "El Celler de Can Roca."

Zephyr Glassware

How can a product, by rethinking its packaging, be introduced into a new scope of use? That was the main focus when designing a packaging system for breakable glassware intended to be used on boats. Zephyr, a premium glassware manufacturer, urged for a solution where sailors could be offered an alternative to plastic glassware. By crafting a reusable packaging that could withstand bounce from a rough sea, and by integrating a multifunctional tray that could reduce the risk of glassware slipping and breaking on a boat, Zephyr was able to achieve the goal.

ZEPHYR OFFSHORE GLASSWARE

SINCE 1892

7 oz   PREMIUM GLASSWARE BY ZEPHYR   20 cl

COGNAC GLASSES

SHORT STEM BALLON SNIFTER

HANDMADE IN THE SWEDISH ARCHIPELLAGO   SET OF

The Zephyr Offshore Glassware is handcrafted and designed in the Swedish archipellago. These unique cognac glasses comes with a tray that allows you to safely store the glassware

02

Gawatt Emotions

Design Agency: Backbone Branding

How do you feel today? Sad or happy? Excited or surprised? Gawatt Emotions cups can say everything for the consumers. In addition to the main identity for the Gawatt coffee shop, Backbone Branding had a task of creating a limited series of souvenir cups. Customers can change the face expression of their cup by turning the exterior sleeve.

Design: Marsel Sheishenov, Kanat Karapashov, Nurbek Nasyranbekov, Nargiza Kulataeva

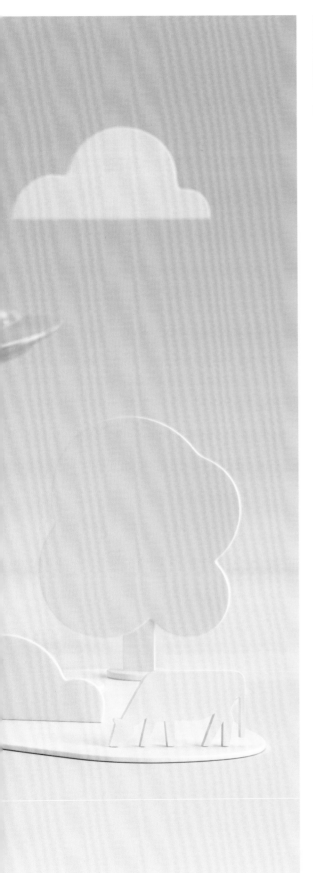

I-Media Creative Bureau's goal is to come up with a milk package that is unlike anything on the market shelves. Molocow is a fun concept package for milk which is designed to look like a UFO to attract kids. The design of the bottle is a tribute to the outstanding director Stanley Kubrick. Kubrick's science fiction film *Space Odyssey* depicts how a cow is abducted by a UFO, which later has become a very popular image in pop culture. This milk package is much like a reconstruction of the classic scene.

Snow Dome

A spray-type room fragrance in the shape of a snow dome. The dome can be lifted from its base, which doubles as its cap, revealing the pump mechanism. Each dome is filled with flowers, leaves, and other tiny forms in polyester film indicating its fragrance. This is a design that disappears beautifully into the room when it is not in use.

Design Agency: nendo | **Client:** by | n | **Photography:** Akihiro Yoshida

Fandango

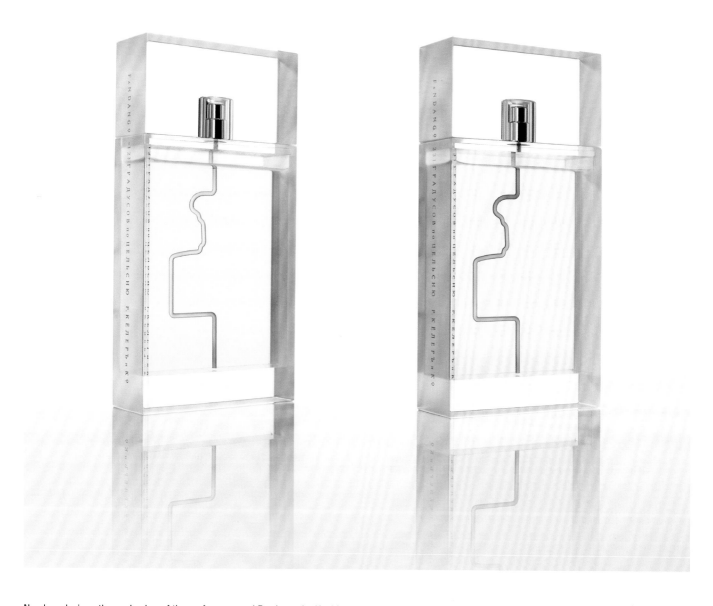

Nendo redesigns the packaging of the perfume named Fandango by Koehler which has gained popularity in Russia in the 19th century. However apart from the image of the perfume bottle, hardly any information on neither the fragrance nor the background to the product is available. This redesign project is to determine the outline of the perfume bottle and express its silhouette with the tube inside. The tubes come in two types, one is a gradation of cold colors and the other warm colors. The perfume bottles are respectively named "Fandango -12.3 °C" and "Fandango +23.1 °C," to suggest the average temperatures of the winters and summers in Moscow where Koehler used to be. These two colors also represent the cool, fresh scent of the winters and the warm, passionate scent of the summers.

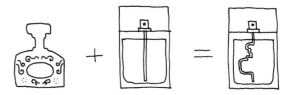

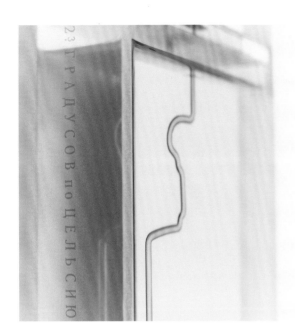

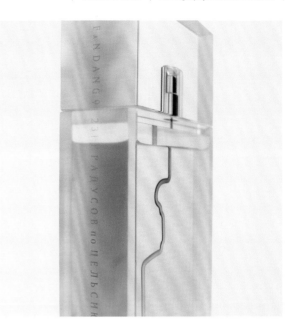

Naked

The most gentle and delicate cosmetics are intimate care products. Naked is a concept design of cosmetics created by Stas Neretin. Its tender color and soft curves resemble the naked body. Every touch makes a tube, a cup or a bottle a little confused. As soon as users take the product in their hands, it will timidly glow right where users are touching. This flirty effect is made possible by covering the package with special thermochromic paint which reacts to warm hands.

Be gentle with this package, it is very shy.

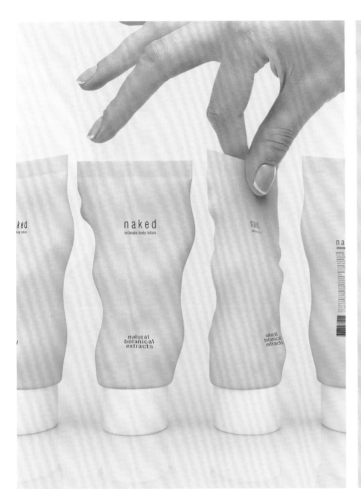

Kif Kif

Design: Matthieu Jeanson | **Photography:** Matthieu Jeanson

Kif Kif is a cannabis smoking packaging design. Matthieu Jeanson conceived this project as part of an academic exercise to produce the packaging for future products. This design allows a perfect preservation of the product by protecting it from light and moisture. Kif Kif refers to "kif" which is the name of cannabis in Morocco, the world's largest producer. "Kif-kif" is a term used in everyday language to talk about two things that are similar. Thus, the designer created a package that could be assimilated to kitchen spices and be introduced into people's environment.

Olive by Gino's Garden

Design: Marios Karystios | Photography: Roger Moukarzel

Gino's garden organic olive oil comes from the region of Rihaneh in Lebanon. Each year, a limited quantity of olive oil is handpicked and cold pressed within six hours to ensure that customers could get the finest quality of olive oil from the grove of Gino Haddad. In order to reflect the organic approach and limited production, a custom made bottle was specially designed to portray the uniqueness of Gino's olive oil. Two different types of uneven olive shaped bottles were drawn, as the designer used mathematical techniques to calculate the irregular shape and volume with the help from Christina Laouri. The ceramic bottles were brought to life by the skilful hands of Stelios Laskaris, in black gloss and green matte glaze finishes. This olive oil is designed in Cyprus, produced in Greece, and enjoyed in Lebanon.

3 Island Gin

This project was a concept proposal for Mc Ewen distillery in the isle of Muck, Scotland, where the gin was produced. The name "3 Island Gin" is inspired from the collaboration between the three small islands Rum, Muck, and Eigg to produce a bespoke gin. The notion of collaboration and sharing came into the design of a refillable and durable bottle which reinforced relationships between the distillery and the customer.

It is a product targeted for adventurous and passionate gin lovers who want to bring forward the thrill of discovery and risk-taking. 3 Island Gin is the best companion for the greatest moments.

Edelbrand Series

Design Agency: Espacioblanco

Edelbrand Series is the packaging design for Vienna Craft Distillery. Every bottle has been dyed by hand for an embodiment of natural fruit's clean and pure essence.

Spine Vodka

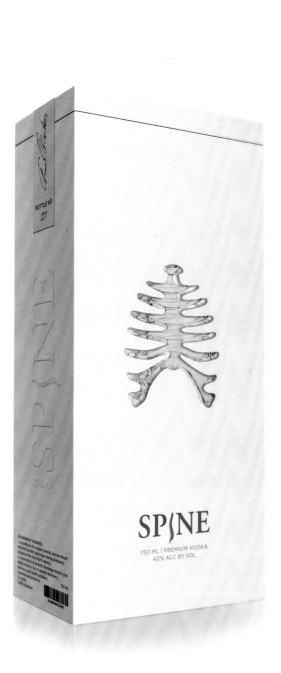

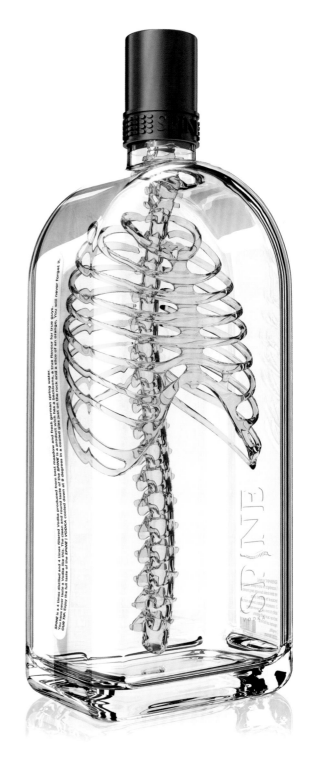

Spine Vodka is a concept for a high quality product. Johannes Schulz applies reduced and simple design with a consciously "twist" in the message. It is created to fascinate the beholder. The integrated eye catching spine with the ribcage communicates a product with a "backbone," which symbolizes a beverage that lives up to its promise to be an honest and pure drink. It does not have to hide anything.

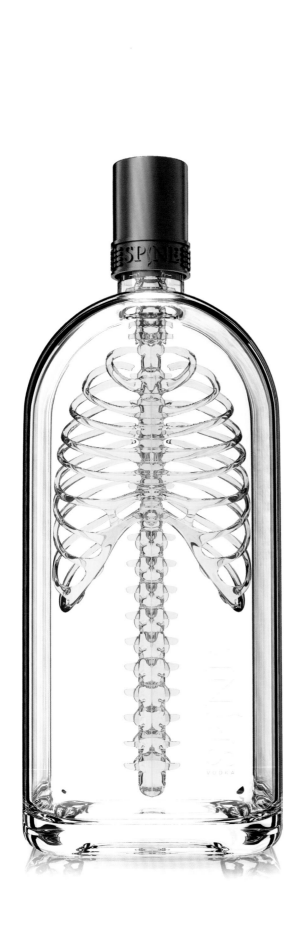

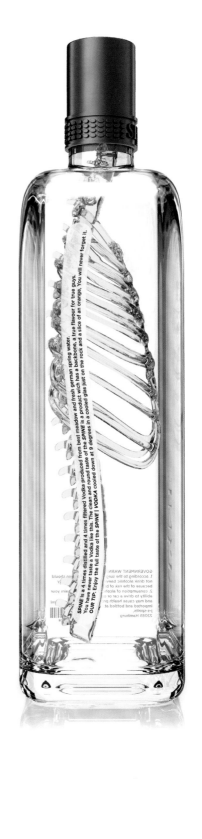

Champlur

Design: Sophie Giraldeau | **Photography:** Mathilde Noblet

Champlur is a bottle of champagne made in the class of
Sylvain Allard, lecturer of the course Packaging Design at
the University of Quebec in Montreal (UQAM). The project
consists in the rediscovery of an object and looks for a
new feature to create a luxury product packaging. A valve
system borrowed from plumbing components is designed
to be the plug of the bottle. This part makes it easier to
open the bottle and therefore bringing the experience of
the high quality product.

Champlur, une expérience des plus
distinguée. Élaboré en quatités limitées
et uniquement dans les millésimes d'exception,
le blanc de chardonnay de cette maison
exprime la noblesse des meilleurs crus de la
Côte des Blancs. Il peut être tout autant
un apéro sophistiqué qu'attablé pour une
expérience gastronomique.

OLIO D'OLIVA

estravergine
senza gocce

OLIO D'OLIVA ESTRAVERGINE
first class cold pressed olive oil

75 ML.
8.5 fl.oz.

produced and bottling in Via N.
Sauro 2. Montale R. Modena

CONSUMIR PREFERIBLEMENTE
ANTES DEL / BEST BEFORE

CONSERVAR EN NEVERA UNA
VEZ ABIERTO / TO PRESERVE IN
THE FRIDGE ONCE OPENED

426254 105728

OLIO
OLIVA

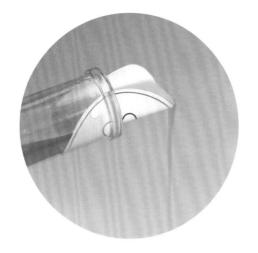

The motivation behind this resourceful vessel comes from the new laws in southern Europe that forbids the use of refillable oil bottles at public eateries. The solution is to add the sticker right on the front of the sleek packaging. The circular section of the label can be peeled off easily and curled into the mouth of the glass neck. This has the double functions — the round window exposes the beautiful color of the oil and reveals how much of the product is left in the bottle. When the bottle is emptied, the entire bottle can be recycled. It is a practical and economical solution for restaurants.

Soligea

| **Design Agency:** mousegraphics | **Creative Direction:** Greg Tsaknakis | **Client:** Troupis Markelou Ltd.

Troupis Markelou seeks for the right packaging for their oil of specific provenance, found between the ancient sites of Corinth and Epidaurus in Peloponnese. Since antiquity this place is famous for its aromatic pinelands that nurture fine olive fruits. Mousegraphics created an unusual bottle, and dressed it in elegant, total, and non-transparent white. The designers laid the topography of the exact area on the bottle surface by drawing its hypsometric curves. This kind of "mapping" of the product avoids all trivial iconography of provenance, creating the fine print of a specific morphology, and suggesting the secret location in a gourmet treasure hunt.

Buteli® Extra Virgin Olive Oil

| **Design Agency:** Chris Trivizas | **Photography:** Math Studio |

Andriotis is a family business, specializing in the trade and processing of olive oil for over 50 years. The characteristic quality of their products is created by a solid know-how and authentic, personal care.

Buteli® is extra virgin olive oil in a small production, which is exclusively produced from olives growing in the magnificent olive groves of Corfu. In order to improve the crop quality, the olive trees are pruned when necessary. The designer collects the branches and turns them into caps for the olive oil packaging. The color of the wood contrasts perfectly with the black bottle, adding a touch of high quality and elegance to the product.

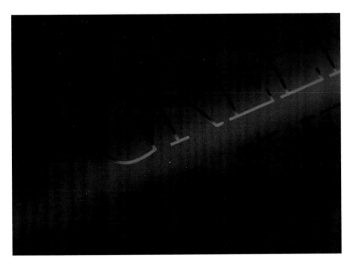
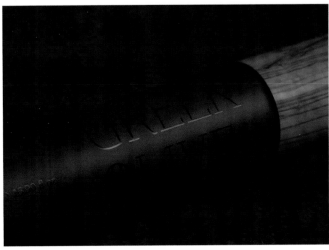

Librottiglia

| **Design Agency:** Reverse Innovation | **Art Direction:** Mirco Onesti | **Design:** Michela De Nicolis |

Librottiglia is where great wine and literary pleasure meet. The name comes from the union of two Italian words: libro (book) and bottiglia (bottle). Three authors, three short stories, three literary genres, and three small bottles of wine integrate into one product. The label is reimagined as a "mini-book" to be read. Consumers can read the stories while enjoying two glasses of fine wine. The stories have been specially selected to complement the sensory impressions of each wine. Textured paper and cord, intriguing illustration, and embossing and UV varnish enhance the tactile aspect of the experience. The result is a perfectly balanced winy-literary experience designed to appeal to a discerning consumer.

| **Relationship Coordination:** Paula Acosta | **Client:** Matteo Correggia | **Photography:** Francesco Zanet, Studio Effe |

Rice Wine

Design Agency: Rice Creative | **Creative Direction:** Joshua Breidenbach, Chi-An De Leo

Rice Wine is a self-promotional product released by Rice Creative agency as a gift to clients during Tết, the Lunar New Year in Vietnam. For 200 years, these wines have been made with great care in the traditional way by one family. It is made using simple methods and pure ingredients in D'ran, Vietnam, a quiet village nestled among the cloudy foothills of the Annamite Mountain Range.

Rice Creative aimed to present the product in a modern way, keeping its internal identity which is minimal and focuses on pure white space. A beautiful gradient appears across the 30%, 40%, and 50% range of wines. It is the strongest characteristic of the range with the inspiration to refer to the product as the bottled cloud.

| **Illustration:** Huy Quoc Le | **Production:** Anna Dinh |

Hand Crafted Tonic Water

The brief was to create four vintage labels for a hand-crafted tonic water brand showing four different flavors. Ashley Wallendorf's concept for tackling this brief was inspired by creating label designs that resemble the vintage, Dutch style of Delft, because Delft stands for a rich history of the Dutch people. The designer's goal was to use the Delft style to highlight the rich history of Dutch ancestry in South Africa, as well as the farming, cottage-like, and country-styled lifestyle which are associated with it, and thus with drinking tonic waters with gin.

C10, the First "Architect" Beer

Design Agency: kissmiklos

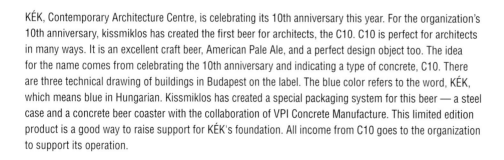

KÉK, Contemporary Architecture Centre, is celebrating its 10th anniversary this year. For the organization's 10th anniversary, kissmiklos has created the first beer for architects, the C10. C10 is perfect for architects in many ways. It is an excellent craft beer, American Pale Ale, and a perfect design object too. The idea for the name comes from celebrating the 10th anniversary and indicating a type of concrete, C10. There are three technical drawing of buildings in Budapest on the label. The blue color refers to the word, KÉK, which means blue in Hungarian. Kissmiklos has created a special packaging system for this beer — a steel case and a concrete beer coaster with the collaboration of VPI Concrete Manufacture. This limited edition product is a good way to raise support for KÉK's foundation. All income from C10 goes to the organization to support its operation.

Lola Beer

Design Agency: Grupo Ingenio | Design: Dani Montesinos

Lola is a special edition beer from the brand Maier born as a tribute to women. Part of the sale went to support the fight against breast cancer.

Die-cut label

Front side

Back side

EGO PREMIUM WATER

Design Agency: Cure Design Agency | **Design:** Bendik Bergh, Brage Brenna-Lund

Every person or athlete with high ambitions needs only the best. To be the best in one's sport, one has to be a person with ego. Cure Design Agency bottled the purest water from the Swiss alps and came up with the Ego premium water targeted mainly men and women with an interest in sports. They are from 25 to 50 years old and have a good income. They own a SUV and are often seen in a colorful technical jacket. To highlight the water's premium quality, designers created a modern and classic package by using monochromatic color and minimalist typography. The brand stands for sporty, healthy, enhanced water, and enhanced lifestyle.

Fragrance FACES

This is a fragrance packaging project inspired by the 1920s which is also called Golden Years. The name "FACES" plays the game of the bilingualism and the difference in the translation. "Faces" in French refers to the multifaceted shape of an object while in English it means the human body part. Fragrance is a luxury item typically anchored in the society and conveys a social status or a lifestyle. Thibault Magni combines the two ideas and designs a fragrance bottle whose shape is inspired by chandelier pearls with geometrical shapes like in *The Great Gatsby*. The visual identity is definitively turned to the old luxury and the box takes a shape of anti-prism.

Secret Sip Whisky

The famous quote by Ernest Hemingway says: "There is no friend as loyal as a book," with a hint of sarcasm, of course. The Hemigway's Secret Sip Whisky is a book-shaped whisky bottle which is finely crafted at the Hemigway Distillery. The Secret Sip Whisky brand is inspired by the famous author Ernest Hemingway with a nice and trendy touch. Secret Sip Whisky is intended for adults who enjoy an occasional sip of finely distilled whisky. The bottle can be placed in a bookshelf if consumers want to keep it as a secret. It is designed to be a rectangular shape for logistical efficiency. When the bottles are stacked side-by-side there is no waste space.

TALAS

Design Agency: ANTI (Oslo, Norway) | **Creative Direction:** Kjetil Wold, Mats Ottdal

For many, drinking beer is about connecting, and re-connecting with friends.

In Hamar, where brewery Basarene is located, "Vi talas!" is an old idiom still in use that can be translated to "We'll talk soon!"

To emphasize Talas' concept, ANTI made an identity built around the very moment when the beers are opened. The typography doubles as openers, which lets Basarene's bartenders open the beers, and it evokes the conversations that follow with Talas' logo. If customers need a little kick-start to the conversation, the packaging paper wrapped around each bottle features stories of Hamar's brewing traditions as well as interviews with local beer connoisseurs.

Ahoy Brewery

Ship Ahoy! Ahoy Brewery is a packaging and brand design for a fictional brand of a micro brewery at the west coast of Sweden. Rasmus Erixon tried to capture the darkness of the story behind the brewery but still got it more interesting and uplifting by using light pastel colors. The design is targeting young adults who knows design and have not had an eye for dark beer earlier.

Ryan Pitman had been brewing craft beer on his own for years. A dream came true when he came across the perfect space to create a brewery in the heart of Minneapolis, on Eastlake Avenue.

Rice Creative soon took note of the rich cultural heritage of the Eastlake Brewery neighborhood, especially a rotating cast of local characters. This inspired the designers to create a set of unique characters which would together represent the community as well as the eclectic brews. Along with these characters, Rice Creative developed a range of logos which hearken back to the olden days of both Minneapolis and Eastlake Avenue, as well as a set of graphic elements which tip a hat to the city of lakes. The cast of working class characters, Minneapolis inspired logos, and unique graphic elements, personify and identify what is at the heart of the brand — finely crafted beers with great character.

| Design: Gregory Jewett | Illustration: TRẦN NGỌC KIÊN |

Chef Tony Bones

Design Agency: bigdog | **Design:** Veej | **Illustration:** Stu Perry

US-based chef Tony Bones, approached the bigdog with the challenge of creating a new identity and packaging for his premium range of barbecue condiments.

The designers named the three products Bone Krust, Bone Kandy, and Bone Fire, and also created the positioning line — Inspired by Fire. Forged by Flavor.

OAK Wine

| **Design Agency:** Grantipo, La Despensa | **Creative Direction:** Sergio Daniel García |

This is an ongoing project between Grantipo and La Despensa for OAK wine.

The idea is to create a bottle made with the same wood of cask used to keep the wine. The wine will be kept in the same environment from the beginning to the end of the process, which means that the cycle of fermentation will not be affected.

Grantipo and La Despensa know that the fermentation in a confined space is different from a barrel and they are studying this process with winemakers.

Hua Diao wine, the fermented yellow rice wine brewed with the ancient methods, adopts the choice glutinous rice and quality wheat leaven, supplemented with clear and pure lake water. It is then stored for a long time to produce a unique flavor and rich nutrition. Through catalysis and brewing, Hua Diao wine resembles the amber in its golden color. The designers liken the wine to the amber which resembles the precious jade when taken from the package. The design reinforces the noble qualities of Hua Diao wine and a sense of dignity. The exquisite traditional Chinese knot reinforces the brand's regional and cultural significance, enriching the overall image and temperament.

Creative collaboration, craft beer, and Christmas are three of Robot Food's favorite things. Collaborating with Vocation Brewery, Robot Food comes up with something naughty, nice, and deliciously festive.

The designers decided on a strong chocolate stout and Vocation perfected the recipe for them to brew together. They came up with the festive name, the elegant in-house illustrations, and chose a special gold foil finish.

Produced in a limited quantity of 1323 750ml bottles, hand labeled, and packed in a bespoke wooden box, it is the perfect Christmas gift for the clients and friends.

| **Creative Direction:** Simon Forster | **Design Direction:** Martin Widdowfield | **Design & Illustration:** Mike Johns |

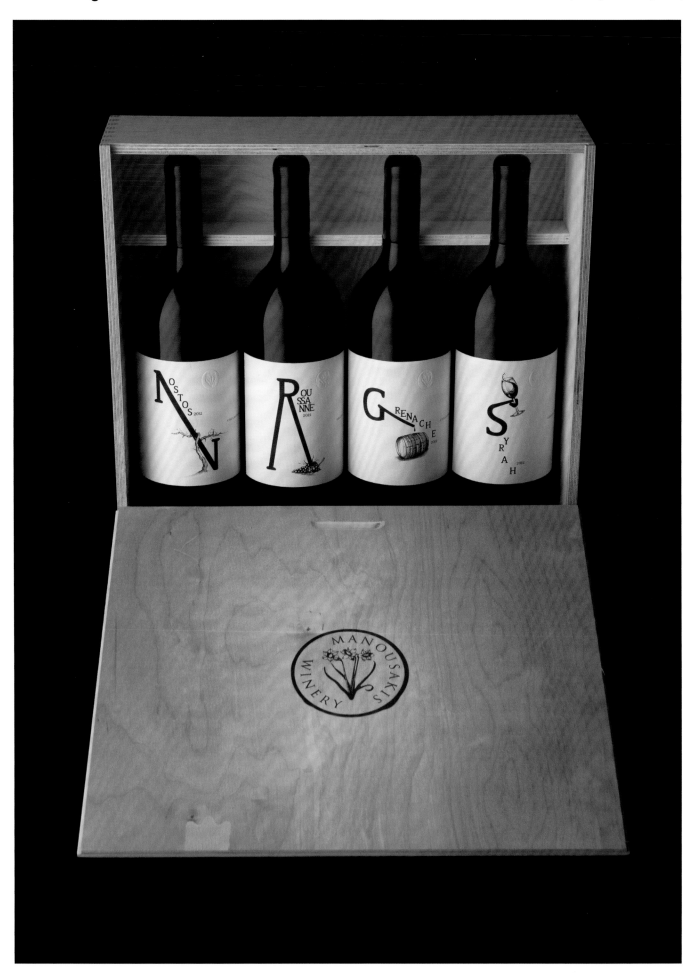

The range of limited edition of magnum wine bottles represent Manousakis Winery's signature wines.

Four stylish scenes are created by playfully mixing typography with pencil drawings that, when displayed in a row, illustrate elegantly the journey from vine to glass.

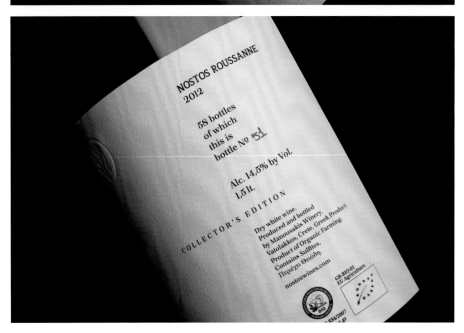

Japanese Sake KOI

| Design Agency: BULLET Inc. | Design & Art Direction: Aya Codama

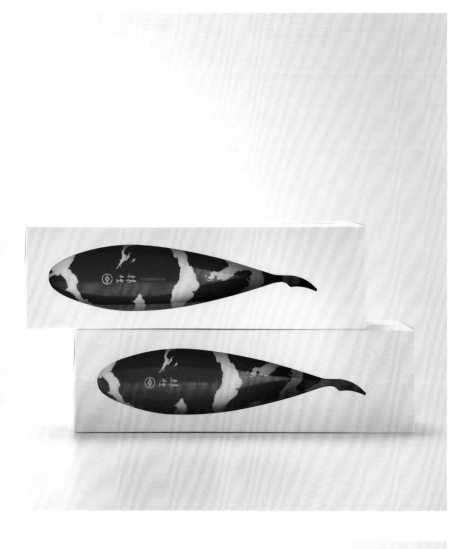

BULLET Inc.'s purpose for this project is to create an impressively crafted sake that represents Japan. The designers choose the famous, ornamental Japanese koi fish as their motif. Koi, called a living jewel, has beautiful red patterns on a white body, and it is attracting many fans worldwide. The package expresses the beauty of koi, with red patterns directly printed on white bottle which resembles the shape of koi. By cutting the box in a koi silhouette, it visually emphasizes the image of koi. This vividly designed package will not only stand out in the store, but also be an art piece for decoration.

Day & Night

Design Agency: Backbone Branding | **Art Direction:** Stepan Azaryan

Day & Night is a concept project with a bifacial nature that runs through each of its design elements. The logo signifies the Earth's rotation with its shape symbolizing the change from day to night. The branding concept centers on the restaurant type that serves dinner during the day and by night becomes a bar. The designers intend to convey the duality of the restaurant's management mode through the colors of black and white with beautifully illustrated animals that are often seen in the Arctic and Antarctic.

Design: Stepan Azaryan, Lilit Arshakyan | **Illustration:** Anahit Margaryan

Diablo Alcoholic Energy Drink

Design Agency: Tough Slate Design | **Creative Direction:** Dima Tsapko |

THE TASTE OF YOUR DARK SIDE

Reveal your other side with Diablo drink in a stylish can. And be a Dark ruler of the nightlife in the town.

Art Direction: Roman Davydyuk, Yaroslav Cherkunov | **Design:** Roman Davydyuk, Yaroslav Cherkunov, Oksana Zmorovych

Masters of Bakery Spices Gift Set

Design Agency: Tough Slate Design

Spices play the main role in an epic play called Bakery. Shape of the package resembles theater cabinet with the posters on it. However, the designers make some adjustments with an attempt to make the theater plays look tasty and sweet.

Art Direction & Design: Dima Tsapko, Roman Davydyuk | **Client:** New Products Group of Companies |

THE INCREDIBLE
MASTERS of **B·A·K·E·R·Y**
PRESENTS
WAITING FOR
CUPCAKE
SPRINKLES | NET WT. 84 G

THE INCREDIBLE
MASTERS of **B·A·K·E·R·Y**
PRESENTS
CYRANO
— DE —
APPLE PIE
VANILLA POWDER | NET 75 G

THE INCREDIBLE
MASTERS of **B·A·K·E·R·Y**
PRESENTS
BROWNIE
— AND —
THE BEAST
COCOA POWDER | NET 98 G

THE INCREDIBLE
MASTERS of **B·A·K·E·R·Y**
PRESENTS
THE
STRUDEL
ORCHARD
CINNAMON | NET 30 G

Mintcraft Electrical Connectors

Mintcraft is an electronics manufacturer specializing in wire connectors and terminals. In redesigning Mintcraft's packaging system, the primary concern was the accessibility of specific connectors for usage. Connector classification was revamped entirely — size is denoted by the height of the package, while wattage and amp ratings are displayed on container lids. These elements are accentuated by a custom numeral set developed in concert with Mintcraft's graphic identity. Connector icons are displayed atop each package for maximum readability when stored. Mintcraft packaging is robust enough to travel in toolboxes, and doubles as reusable storage for additional connectors purchased in the future.

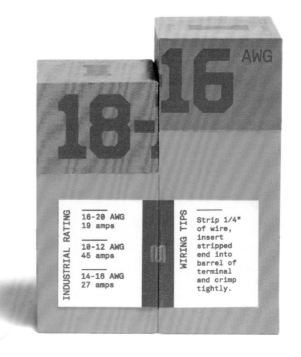

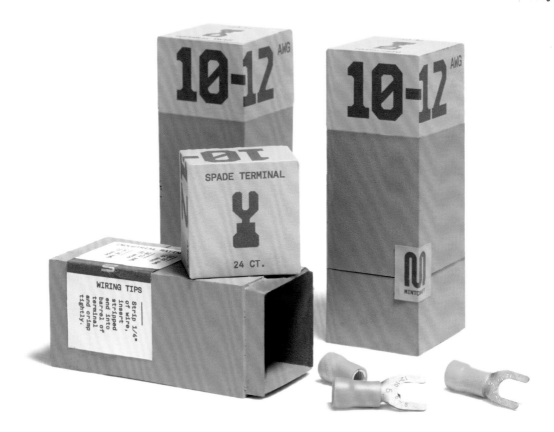

Passkey

Design: María López Benítez

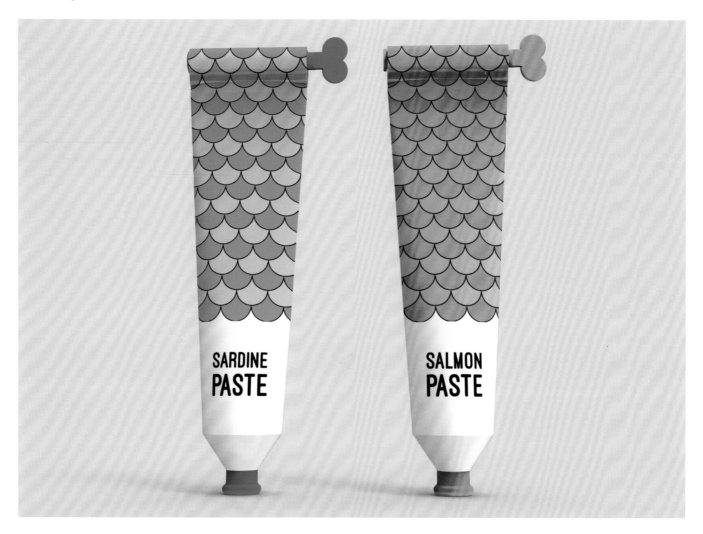

Passkey draws inspiration from the ancient keys used for opening cans. The designer maximizes the content of the package by adding a hole in the end of the aluminum tube where a key could be inserted and twisted every time the product is consumed.

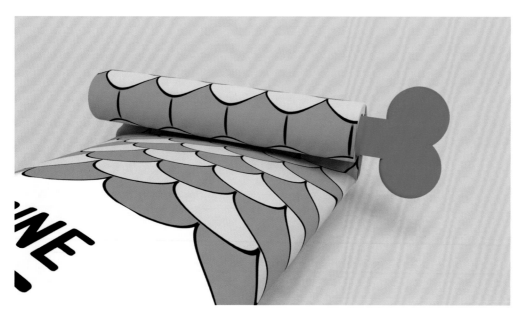

Cereal for Adults

| Design: Mun Joo Jane |

The overview of this project was to come up with a cereal packaging design for adults. The challenge was to neglect the stereotypical cereal box design. The brand Special K was chosen for this re-branding project. The designer had an urge to change the Special K into a unisex design, leading to the creation of a b-line for Special K called "K+." "K+" creates a positive reputation and symbol for the packaging.

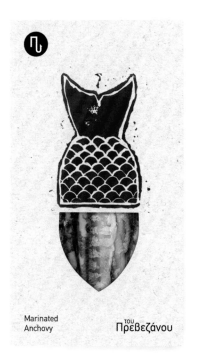

Marinated
Anchovy

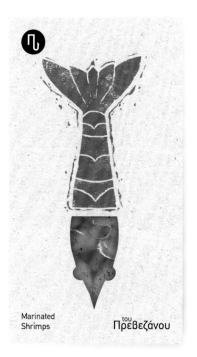

Marinated
Shrimps

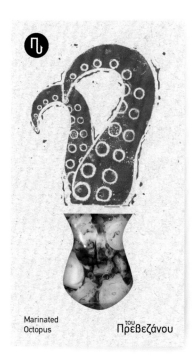

Marinated
Octopus

Tou Prevezanou is a new series of frozen and pre-cooked fish and crustaceans launched at the Greek market by Ionian Fish, a family business from Preveza. For the new products' branding, 2yolk "fished" for its inspiration from the long fishing tradition of the Goumas family.

2yolk decided that the Prevezanos packaging should have the simplicity and freshness of a good fish delicacy. A simple die cut on each package allows the consumer to see the quality of the product, while the fish are illustrated using the linocut printmaking technique. The result, a handmade, almost rough look, captures the manual, delicate production of these delicacies, as the Goumas family tradition dictates.

Daily Affections

Design: Albert Junghwan Son

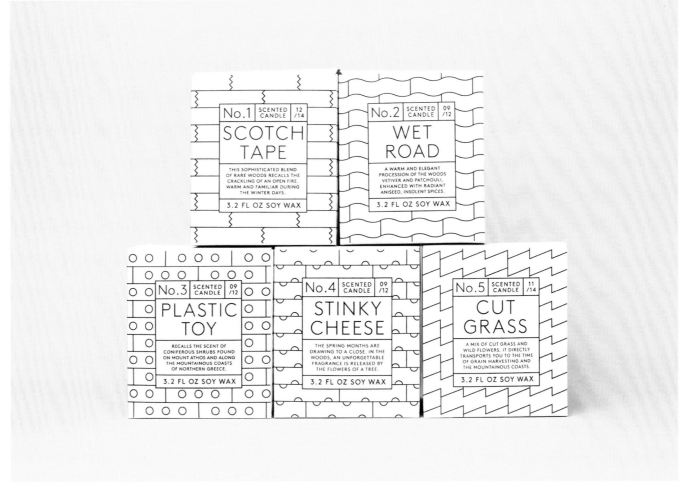

Daily Affections is a series of candles featuring odd but attractive scents we encounter in our daily lives. The designer uses simple lines and dots to create five clean patterns that symbolize the scenes of everyday life and signify the different scents of the candles.

Tesis

| **Design Agency:** Anagrama |

Tesis was born in Mexico, under the curatorship of Sommelier Marcela Garza. Its unique herbal and tea blends are inspired by the origins of tea and ancestral heritage from flowers, herbs, roots, and fruits. Japanese calligraphy, in which lines and dots become remarkably important, represents one of the most popular arts in Japan.

Inspired mainly on the Japanese Art, Anagrama utilizes water and ink stains to represent the complexness and lightness of tea mixtures. In the same way, vertical typographic arrangements based on traditional Japanese reading are employed, obtaining a balance between classic and modern typographical styles. The main emblem symbolizes the Japanese wind chimes "Fuurin" placed on doors and windows of homes in early summer. The color palette focuses on natural tones with red color accenting details and gold foil rendering elegance.

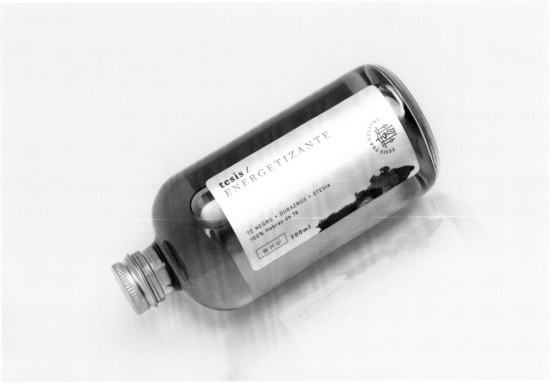

Krasnogorie

Design Agency: Science Agency | **Art Direction:** Pavel Konyukov

Krasnogorie (Red Mountains) is a Ural brand of meat delicacies and sausages from the Chelyabinsk meat processing plant, which dedicates to providing regular customers with the truly high-quality, natural sausages made from actual Ural meat.

By utilizing the technique of linocut and making a perfect combination with the colors red, white, and graphite, Science Agency took the vivid image of a juicy steak as a basis for the brand identity to underpin the nature of the products.

Such "meaty" background is easy to discern from a distance. Upon a closer view, the pattern of meat fibers takes the shape of farms, villages, Ural Mountains, and free-ranging cattle, which connects the packaging to the brand story and speaks not only of the product's quality but also of the producer's approach.

Layer Hen

Design Agency: Geometry Global (Russia) | **Creative Direction:** Andrew Ushakov |

Layer Hen is a business souvenir in a form of a limited-edition egg carton. The designers combined the package for a dozen eggs with a grocery bag, and decorated the carton in the form of a hen.

The package was aimed to demonstrate how fresh the eggs are, and how carefully each of them is produced.

Otoño is 100% Ibérico Spanish ham made from free-range pigs raised in a sustainable environment. Tres Tipos Gráficos chooses the brand name, "Otoño" (Autumn), as it represents a direct reference to the idea of climate and natural environment, as well as to the trees that produce the acorns, which are crucial to the quality and distinction of this exquisite product. The packaging collection functions as a set, where each of the three main products is represented with its own tree species. The leaf blueprints provide each product with its own individual identity, thereby enriching the brand experience of the whole product range and creating a strong link to the brand name "Autumn."

Got One!! Wild Mullet Catch

Design & Illustration: Ching Wei Liu

Mullet fish roe is famous as a true delicacy. The packaging design with its premium appearance focuses mainly on the use of black as a color to visualize the fact that this fish is named for its black backside. In addition, the color concept aims to picture the specific atmosphere of the fishing season when the coastal fishing villages are then covered with the golden light of mullet fish roe that have been laid out to dry. For this reason, mullet roe is known as the black gold of the sea.

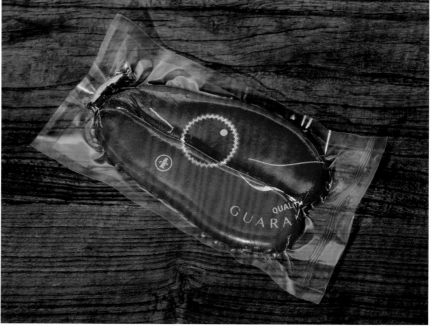

SCW Salzberg Chocolate Werks

This is a packaging design project for the 100th anniversary of the world's renowned chocolatier, Salzberg Chocolate Werk. The design direction is inspired by Wiener Werkstätte and artists from the Vienna Secession. To provide a collectible and high-end gift box, the designer creates three modular metal cans and a refined wooden box for the package, coupled with gold foil logos. The design conveys not only the artistic history of the brand but modernist design style through the combination of various materials and the graphic details shown on the labels.

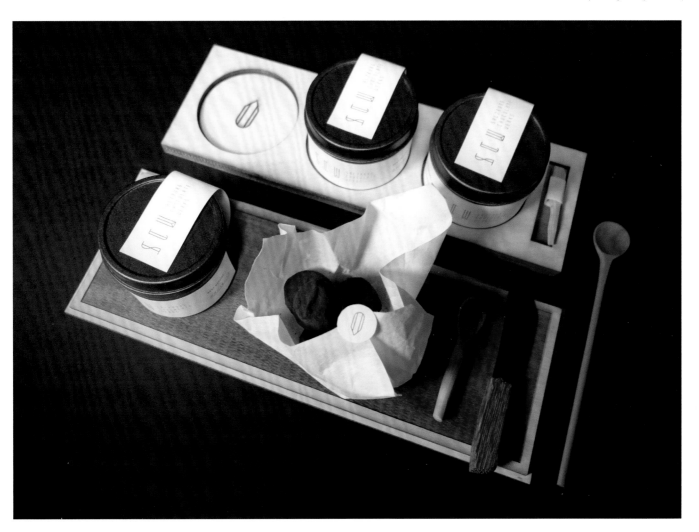

Caffè Pagani

Design Agency: Eskimo design studio | Art Direction: Pavel Emelyanov

Roasting Caffè Pagani, based in Lugano, was founded in 1949 from the dream and the passion of the founder Giorgio Pagani. The new logo is a metaphor of circular movement of coffee during roasting. Everything is circular and everything rotates. Generation by generation Pagani offer customers coffee of excellent quality. Eskimo design studio drew inspiration from the device of roasting machine for the packaging design. Usually a roaster has cylinder form and special round window on the front side to see the degree of roast. The designers used these typical details in packaging concept to convey the idea of fresh roasted coffee. Each coffee has its own name and capital letter for quick identification, for instance, "E" for Espresso, "G" for Grani Naturali, etc. On the back side of packaging there are circular messages for real coffee lovers.

| **Design:** Irina Emelyanova | **Photography:** Anatoly Vasiliev |

A Bowl of Rice

Design Agency: Shanghai Version Design

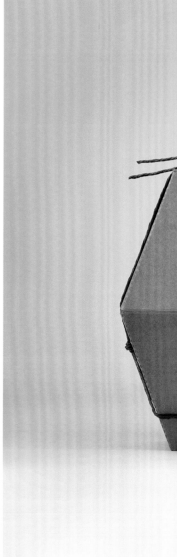

A Bowl of Rice is the package design of organic rice. This concept takes inspiration from the traditional Chinese rice container in countryside. The designer uses the new structure to interpret the traditional concept, making "food" and "container" more interesting in daily life.

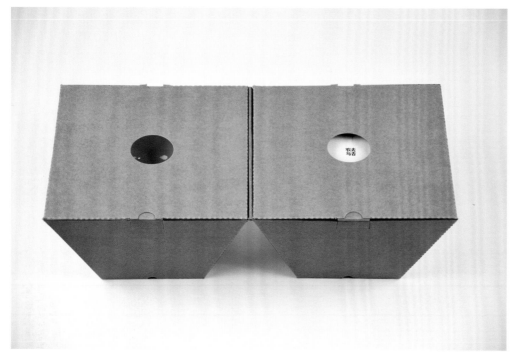

MØS Gastronomic Smart & Casual

Design Agency: Backbone Branding | **Art Direction:** Stepan Azaryan

MØS Gastronomic Smart & Casual is a Nordic restaurant in Moscow presenting the Scandinavian cuisine, traditions, and philosophy. Backbone Branding dived into the Scandinavian philosophy and caught the essential of the brand — minimalism and simplicity. MØS restaurant celebrates the Nordic style and becomes a strong brand with its own traditions that values the nature and healthy lifestyle.

Galamb Tailoring identity

Design Agency: DekoRatio Design

Galamb Tailoring offers the highest quality handmade suits in Budapest. "Galamb" means "dove" in Hungarian, a name derived from their previous location on Galamb Street. The overall rebranding and visual identity of Galamb Tailoring reflects the key elements of the brand — style, perfection, and tradition.

The design was aimed at capturing the brand's essence in a simple, elegant yet instantly recognizable way. The symbol was composed of a jacket collar and lapels, which resembled a flying dove and resonated with the idea of bespoke tailoring.

Adidas Athletics Z.N.E Hoodie Launch

| **Design Agency:** Colt | **Creative Direction:** John Sweeney |

To announce the arrival of Adidas Athletics, Colt was briefed to create premium packaging solutions for one of the largest cross category activations that the brand had ever seen. The Z.N.E (zero negative energy) hoodie is designed to block out any distractions and create a zone around athletes where they can find focus. With the direction of the product in mind, Colt began working on a raft of packaging concepts that would create a protective "zone" around the contents. Colt ultimately agreed on the soundproof concept as it really got to the heart of the brand and had strong visual and textural ties with materials being used across the range.

Design: Alex Chappell | **Client:** Adidas | **Photography:** Michael Hedge

Klässbols Linen Factory

| Design: Rasmus Erixon, Tobias Möller |

| **Client:** Klässbols Linneväveri | **Photography:** Rasmus Erixon, Tobias Möller |

Klässbols linen factory is a family owned company established in 1920 by Hjalmar Johansson. The brand has a long and genuine history and is one of the few linen factories that still exists in Sweden. Klässbols produces products of high quality by working with designers to create modern and challenging patterns. They value highly the tradition of handicraft and craftsmanship. This design is to convey the brand's craft through modern design by developing the packaging solutions that reflect the essence of their products.

CS Electric

Art Direction & Design: Angelina Pischikova | **Illustration:** Anna Orlovskaya

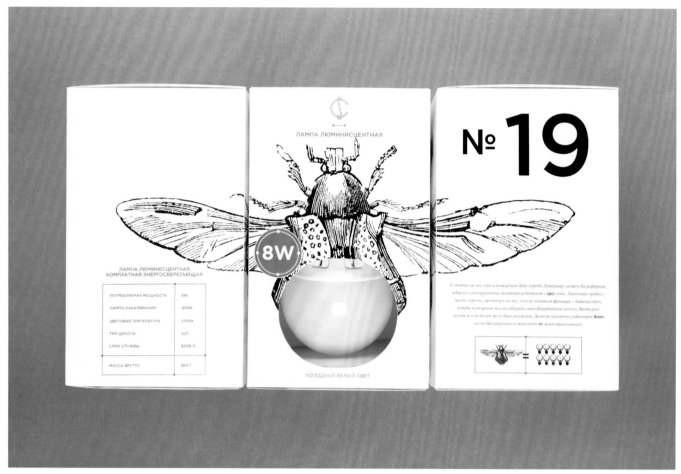

CS is the largest electrical company in Belarus, which supplies the market with more than 5000 items. CS company's branding and packaging is inspired by the old physics books that feature diagrams of electrical circuits and illustrations of scientific experiments. Inspired by Thomas Edison's words that a firefly is an ideal cold light source, the designers came up with the idea of comparing the various forms of light bulbs with various insects.

The detailed illustrations show each insect with its wings spread wide, allowing the light bulb body to be seen. Each bulb has its own shape, from long and skinny to stout and fat, adding a bit of personality to each bug illustration. The wattage on the front allows consumers to choose the one best suits their needs.

This is the branding and packaging design for Marine. The design inspiration comes from its own products. The distinguished features of each fish are brought forward by the bold design and realistic texture, which also highlights the freshness of the products.

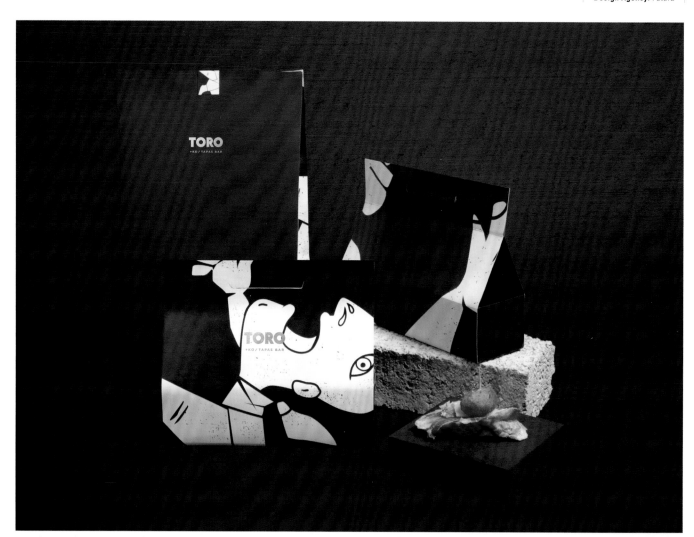

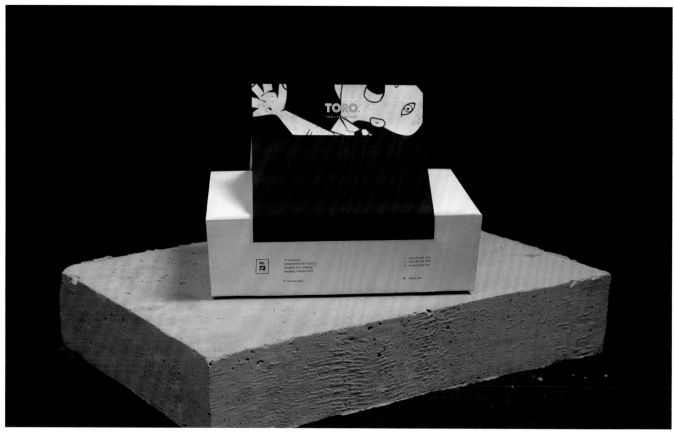

TORO is not a traditional Spanish restaurant for old tourists, but a cool place for people who want to have a great time and enjoy amazing food and great wine. With this in mind, Futura created the branding and packaging centered on a more modern style with black and white as the color scheme. To communicate the Spanish vibe without being too obvious, they took some elements of Spanish culture for inspiration. On one hand they did a reinterpretation of Picasso's Guernica mixed with one of the most iconic festivals in Spain, namely, San Fermín. On the other, they drew inspiration from Gaudí's mosaic technique, that is, Trencadis.

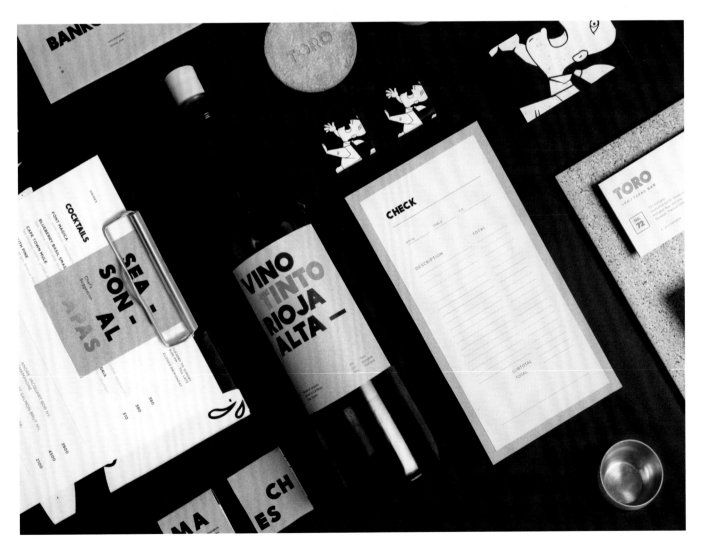

Madame Quoi

Design: Sofia Villarreal Samperio, Cynthia Fernandez Arellano

Madame Quoi is the branding and packaging for a gourmet bakery inspired by the dramatic French story of Henriette Caillaux set in the year of 1914. The story is about the woman Henriette, a Parisian socialite and wife of a French politician, who kills the editor of a newspaper in order to stop him from publishing a love letter between her husband and his mistress. This project seeks to reflect the explosion of feelings of the heroine. The misalignment of emotions is reflected throughout the use of experimental textures. This typical graphic language for the brand is settled and lives through the different packaging, creating a distinctive, modern, and unique brand system.

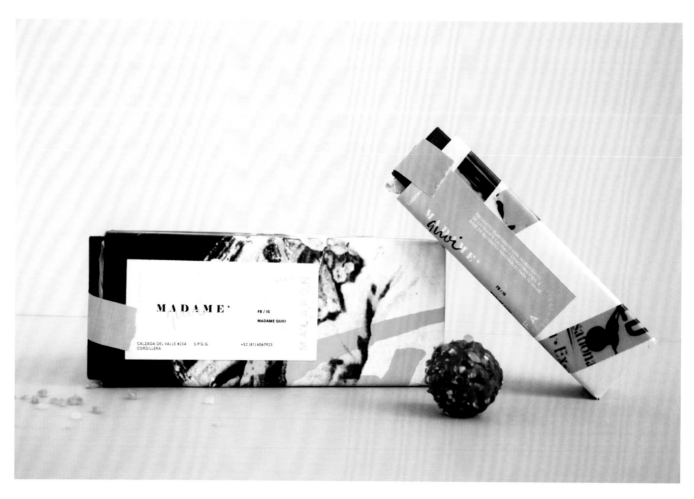

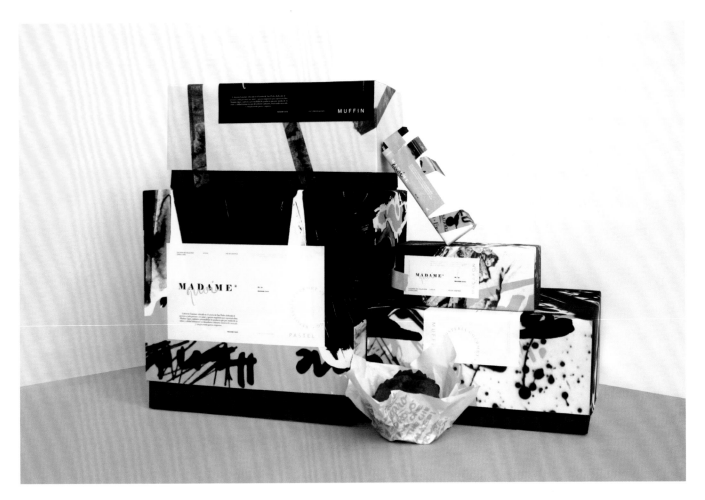

FICO is a homegrown brand of ginger ale guaranteed to have its flavor worked its way to the consumer's heart. This ginger mixture is packed with all natural ingredients that is lovingly brewed into each bottle by a dedicated team of "ale-o-holics." Latching onto the brand's ideals of transparency, Shao translates this element to its branding and packaging as well. The logo shows what is inside every bottle with a shift in the colors of the wheel. The packaging itself shows the story behind the brand so customers can see that FICO is made fresh from the farm, to the bottles they now hold.

Juiceline is a cold pressed juice brand and bar. Firstly kissmiklos focused on the juice bottles and the juices colors. The logo points out the broad range of the juices available. It consists of a variable number of Juiceline bottles on a shelf and a simple logotype of the brand. This minimalistic design highlights the natural colors of the juices. The menu card has been designed by analogy with the Pantone color chart and the numbers on the label are real Pantone codes.

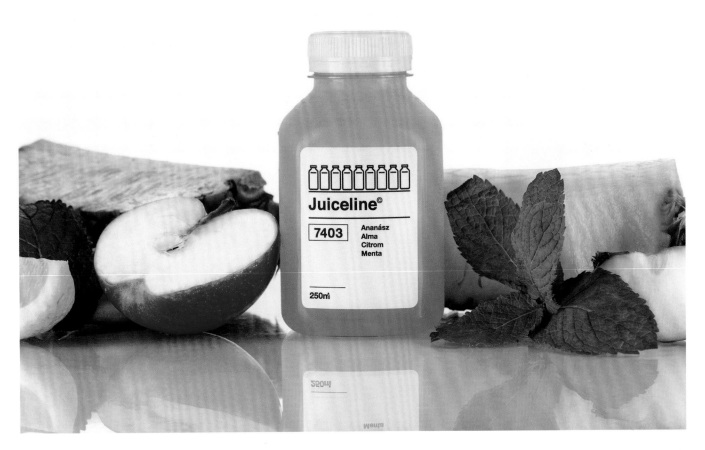

N'1CE Cocktail, the ice cold and frozen cocktails, which comes in classic flavors like Daquiri, Pina Colada, Mojito, and Margarita, is made to enjoy on warm and sultry summer evenings as well as intimate clubs. N'1CE is designed to be an easy-to-grab tube and served frozen, which is not only refreshing but prevents drinks being spilled everywhere.

Bright and electric colors are used in the packaging to avoid looking too youthful. Each flavor has its own color palette inspired by the original cocktail, visualized by using acrylic paint on glass. The packaging architecture is based on four elements: the background pattern symbolizing the flavor, the stamp highlighting the 5% volume alcohol content, the product name, and the signature of award-winning bartender Jimmy Dymott which ensures the quality.

Design: Antonio Vergara Alvarez, Mia Askerstam Nee | **Client:** N1CE Company

Savon Stories Lotion Bars

Design Agency: Menta | **Art Direction:** Laura Méndez

Savon Stories is an English company specialized in the handcraft of 100% organic soaps produced in small batches, through a cold-processed method that retains all nourishing properties to feed the skin. Savon Stories Lotion Bars is their new luxury line made with raw ingredients and essential oils to nourish the skin deeply. Each wrapping features a botanic illustration which indicates the main ingredient of the Lotion Bars.

| **Design:** Margaux Sarlin, Blanca Jiménez | **Client:** Savon Stories | **Photography:** Paola Chávez |

The National Gallery Singapore, a new art museum, was keen to partner with Marou Chocolate and co-brand a range of chocolate bars for their store, which would represent both Vietnam and the exceptional architecture of National Gallery. The Gallery is comprised of three distinct spaces — Historic, Modern, and Transcendent. Rice Creative created a set of icons based on these themes, while Marou conceived three chocolate flavors that matched them in spirit.

To properly represent Vietnam, Rice Creative sought to wrap the bars in an authentic and traditional Vietnamese art form. The clear choice was Đông Hồ, a traditional print-making art form, which originated in a small village in the North of Vietnam. The designers worked with a family who has been practicing this art form for 500 years. Everything in the process was hand made.

Creative Direction: Chi-An De Leo, Joshua Breidenbach, Gregory Jewett | **Design:** William Sorqvist, Anna Tran | **Production:** Anna Dinh

Luk Yik Kee Package

| **Design Agency:** invisible | **Design:** Lam Cheung

Luk Yik Kee is one of the oldest handmade cutlery houses left in Hong Kong. Using mother-of-pearl as the core material, Mr. Luk Si Yik established this unique craft since 1900. It is until 2015 that Mr. Luk's great granddaughter wanted to revive the vintage artistry into a modern brand. The sophisticated history are commemorated in the six sides of the Chinese scroll. The hexagonal design represents the family's name "Luk," which sounds like the number "six" in Cantonese. Through the design, the customers can experience the brand's artistry and appreciate Luk Yik Yee's age-old story.

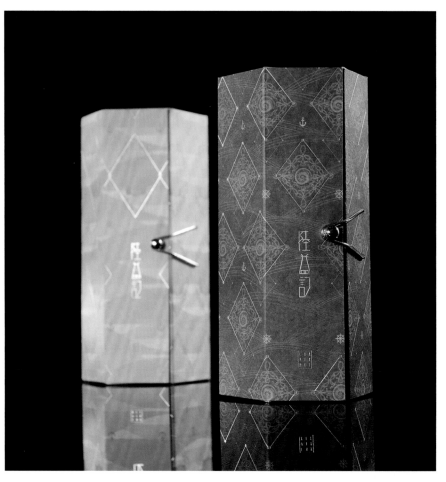

MAKE ME LOVELY CUSHION

Design Agency: TRIANGLE-STUDIO | **Art Direction & Design:** Kisung Jang |

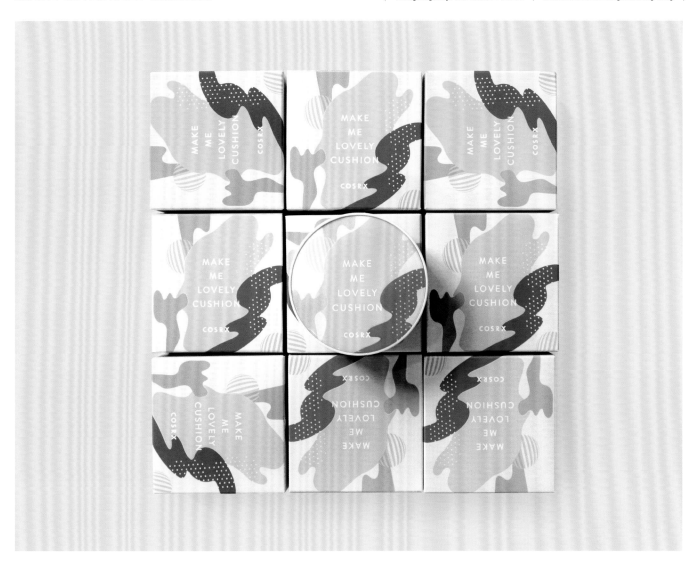

MAKE ME LOVELY CUSHION is a cosmetic product of COSRX. To match the name with the graphics of the products, the designers uses a darling color palette for the brand's cream and cushion with abstract graphics that gives off a soft and sweet vibe. The light pastel colors make the brand youthful and radiant.

Luxury Skin Cells

Design Agency: Lavernia & Cienfuegos | **Client:** ETNIA Cosmetics

Lavernia & Cienfuegos attempted to create the brand identity and product design for ETNIA Cosmetics, which would convey the brand's qualities such as elegance, sophistication, tranquility, and exclusivity. The designers used different finishing methods to create a final look for the brand's premium product. The organic metallic shapes evoke the idea of cells, giving each of the products a unique personality. The text is blind embossed without ink in order to highlight the image of the "cells." The colors and forms of the jars work in concert with the whole range.

BARRO DE COBRE

| Design Agency: Savvy Studio | Photography: Savvy Studio |

Barro de Cobre (BdeC) is a mezcal that believes in taking its time and doing things the proper way. This is why it is distilled twice — once by using a clay pot and the second time round, in a copper pot. BdeC is proud to be the first to conceive this unique blend of different distillation processes, from which its name was drawn — strong yet smooth, clear yet earthy.

Savvy developed a graphic language that reflected the characteristic duality behind BdeC's processes, by overlapping two seemingly opposite textures and finishes. On one hand, they came up with a colder and more industrial side — white satined papers, and on the other a warmer and more artisanal one — uncoated papers and embossing. The brand's logo reflects the slight imperfections caused by hand, tracing an otherwise perfect digital typeface, which once again, represents the opposing yet complementing forces behind this mezcal.

Conlori Contact Lenses

Design Agency: invisible | **Design:** Lam Cheung

This is the packaging, corporate branding and naming design for CONLORI contact lenses. Each package contains only one contact lens, so the users can mix and match it everyday. In search for a word that would represent the playfulness and creativity of the colors of these contact lenses, the studio comes up with the name "conlori" inspired by the Italian word "colori" which refers to color. The designer creates a logo that reflects the whimsical nature of this personalized daily product. A color scheme of black and white adds a touch of modernity while at the same time does not overwhelm the users' freedom to play.

Once A Month – Tampon Packaging

Design: Olivia Chan | Art Direction: Suzanne Cote

A tampon packaging for Once A Month, an organic menstrual care company, who finds importance in the environment as much as the female body. They aim to produce high quality products and present one of the most "taboo" products in a graceful way. A minimalistic approach was applied to create a luxurious feel because many women are still embarrassed to carry such product. Symbols of the moon are used to illustrate the fact that some women are able to synchronize their menstrual cycle. The brand also wishes to create a community among women to share insight and tips for health and environment care. Hence the designer used the number "1" in the front of the packaging to imply that all women are in one big community.

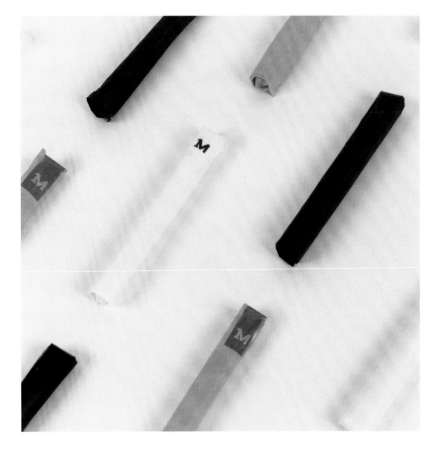

Rond Et Carre

Collaborating with an eyewear manufacturer known for its traditional handmade techniques since the 1980s leads to the creation of "Rond Et Carre." The circular and rectangular shapes of the frame hints the fusion of the two brands in creating a unique design. The circle and the square represent respectively the two brands: MUZIK and PARAGRAPHE. The design prides in balancing the two simple shapes. The end piece is also decorated with square on one side and circular on the other. The packaging is designed to mirror the packaging of gifts which gives out a sense of excitement when it is being unwrapped.

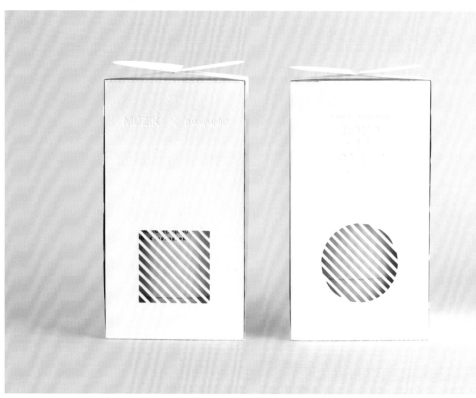

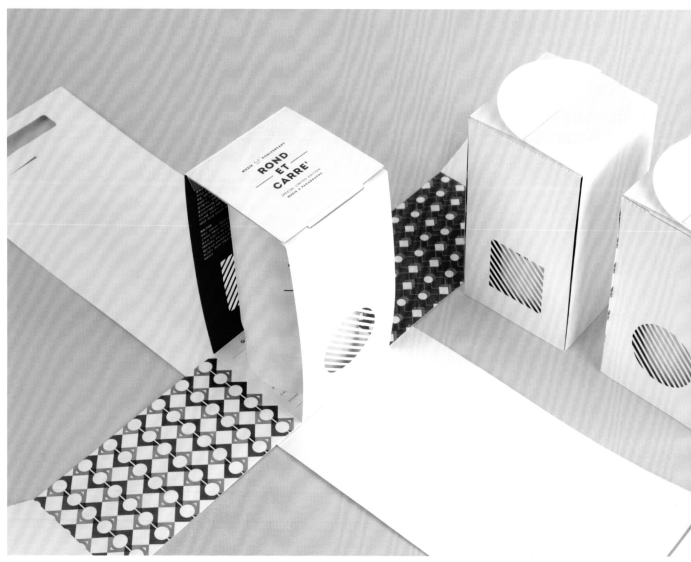

TMA-2 Modular

| **Design Agency:** AIAIAI, Kilo Design | **Creative Direction:** Peter M. Willer |

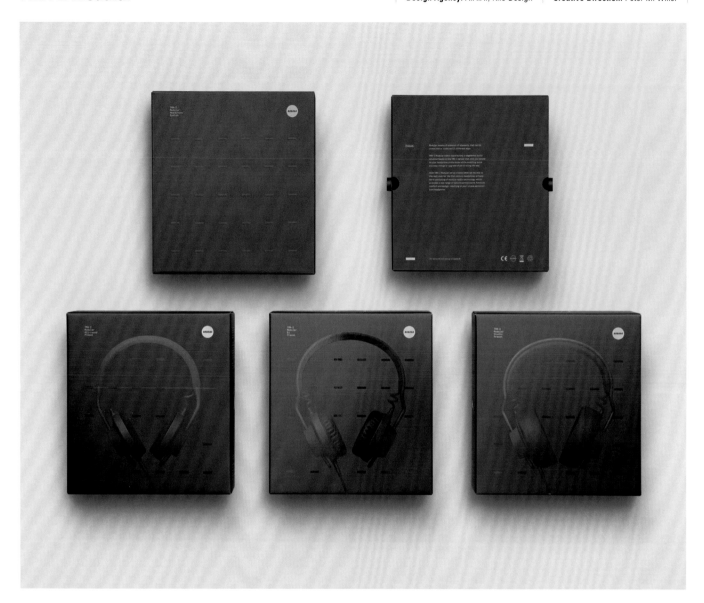

TMA-2 Modular Packaging System is a collaboration between AIAIAI and Kilo Design. AIAIAI needed a packaging scheme for its TMA-2 modular headphone with over 20 parts and more than 500 combinations. They looked for a design that could serve the product delivery and retain the fun of unpacking.

The result turns out to be minimal, functional yet playful with visual grid and pattern. Clean typographic design is applied on both the box and the product part bags. The standard top box has the modular grid embossed along with the screen printed logo and text. This box connects with a lower box part. The two boxes are held together by a plastic peg on each side of the lower one which functions as a lock system. Each box contains four parts in individual bags — headband, speaker units, earpads, and cable. Stickers on the front of the bags showcase the precise details of each parts.

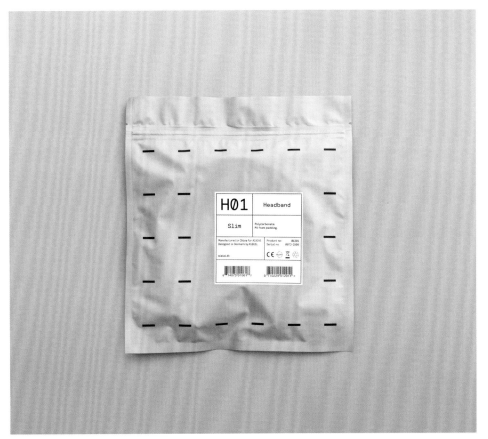

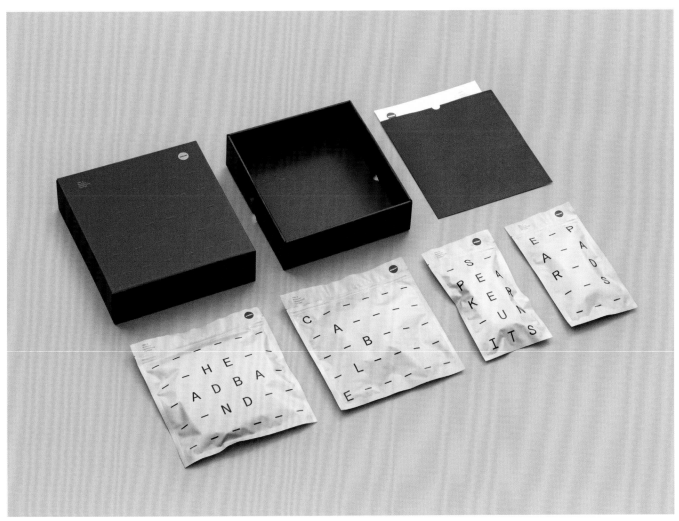

Dark Brew Coffee House

Design Agency: Black + White

VADER

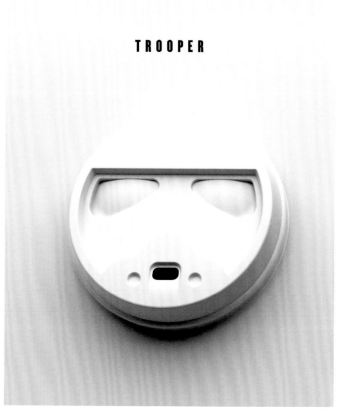

TROOPER

Inspired by the dark side of the force, Dark Brew Coffee is a packaging and branding project with an aim to offer a new level of caffeine experience. This conceptual coffee brand stands out from the rest with its tailored-made coffee cups with images of Darth Vader and Stormtrooper on the lids. The designers also engrave on the cups the modified quotes from the legendary *Star Wars* movie. This brand not only offers an ultimate joy to *Star Wars* fans, but also approaches the mass through coffee bean packaging with stylish design.

Design: Spencer Davis, Scott Schenone

Cito

Cito is a little place located in the always young Colonia Juarez in Mexico City. Local people use the word "cito" as a common name for the small things. In Mexican culture, food portions are large and this often causes overweight. By contrast, Cito provides a controlled menu that cares about people's little heart. Estudio Yeyé created a simple and clean packaging system to work concert with the brand's philosophy.

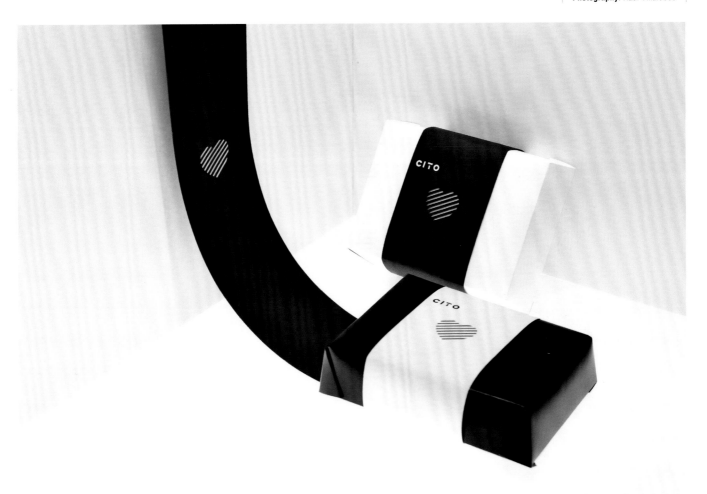

DOSE

| **Design:** Nora Kaszanyi | **Art Direction:** Tamas Marcell |

DOSE is a school project designed by Nora Kaszanyi. She was asked to create an identity for any fictional soft drink brand. The only requirement was to come up with six different flavors and give the bottles a completely different look with distinguishable colors, icons and so on. DOSE shows a flavored soda water crammed with vitamins and minerals which explains the reason why the designer has chosen the pure and medical bottle shape with pale colors. The goal is to create a drink that customers would love to consume after an exhausting long night, like enjoying a cup of coffee without caffeine.

Photography: Sara Szatmari

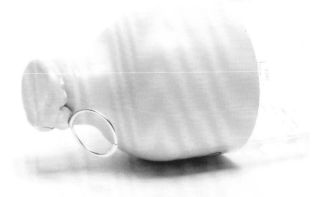

Beatific

| **Design Agency:** mousegraphics | **Creative Direction:** Greg Tsaknakis |

Mousegraphics was commissioned to create a brand story, a design, and an identity which would be able to communicate equally the science and aesthetics through the most refined language. After a research in art history and linguistics, mousegraphics proposed the brand name "Beatific." "Beatific" is an existing term linked to a notion of the "blessed," an inner communion with the divine, and the sharing of its spiritual light, splendor, and glory.

Mousegraphics used "Beatific" in their design as the meeting point of "beauty" and "scientific," and they designed everything on this axis — the abstracted drawing light patterns on the packaging, the use of semi transparent, translucent material, the four color palette of iridescent hues, and the elegant yet serious typography. They worked on this visual language by designing stands, publications, and everything else related to the communication of the product's attributes.

Zee – Honey Goods

| **Design Agency:** Gen Design Studio | **Creative Direction:** Leandro Veloso | **Design:** Catarina Correia

Gen Design Studio portrays Zee's values through the creation of a solid communication strategy, including naming, logo, graphic elements, and packaging,

The peculiar shape of the package clearly illustrates the brand's personality and allows it to stand out from the competing goods. By keeping a simple design aesthetic, the designers carefully place the information on the frontal side to attract instant perception. The entire package is made of translucent paper which is resistant and visually pleasing, creating a distinct and unique identity for the brand.

Game of Thrones, and design with more finesse than Anna Wintour's hair. They love what they do.
P.088-089

BULLET Inc.
bullet-inc.jp

BULLET Inc. is a design firm established in Tokyo by Japanese designer Aya Codama. The firm offers wide-ranging graphic design solutions including packaging, logo, and web design. BULLET Inc. particularly excels in packaging design, with constant attention to the actual texture, weight, and volume of materials to enhance the product's original appeal.
P.158-159

Ching Wei Liu
www.behance.net/robomb-devours

Ching Wei Liu, a.k.a. Devours Bacon, is a Taiwanese designer. Most of his designs do not originate from a briefing. Instead he always starts with an image or a slogan which does not relate to briefing but instincts. His work has been three shortlisted winner for some design awards including Red Dot and Gold Pentawards, etc.
P.180-181

Chris Trivizas
www.christrivizas.gr

Chris Trivizas operates within the field of visual communication since 2003. It is a creative studio committed to providing premium quality, integrated communication solutions tailored to each client's needs. Their services include branding, corporate identity, logotype, stationery, invitations, flyers, brochures design, and packaging.
P.129

Cindy Forster
www.cindyforsterdesign.com

Cindy Forster is a professional art director and designer living and working in Sydney, Australia. Strategic brand thinking, together with a passion for great design and love of the craft, drives her elegant and effective design solutions. Her minimalist approach to aesthetics ensures the thoroughly fresh, modern, and highly considered design.
P.018-019

Claudia Lepesqueur, Maria Vidal, Maria Elipe
www.behance.net/claudia_lepesqueur (Claudia Lepesqueur)
www.behance.net/Mavikka (Maria Vidal)
www.behance.net/mariaelipe (Maria Elipe)

Claudia Lepesqueur, Maria Vidal, and Maria Elipe are graduate students majoring in Packaging Design at ELISAVA University in Barcelona, Spain.
P.084-085

Colt
www.thisiscolt.com

Colt is a British design studio dedicating to creating emotional connection between a brand and its target market through design. They help brands to sell their values and ethos. Colt believes that it is the thinking that makes the difference between "pretty" creative work and powerful creative results.
P.192-193

Communal Creative
communalcreative.co

Communal Creative was founded in the belief that great brands are rooted in good design. Communal Creative are committed to finding unique moments to connect, differentiate, and provide memorable experiences for their clients. They offer a wide range of creative services ranging from brand identity to physical and digital designs.
P.083

Cure Design Agency
www.thecure.no

Cure is a design agency based in Bergen, Norway. They design and develop visual identities and websites for small to medium-sized businesses. They operate on the principle of "no cure — no pay," which means that if a business does not increase competitiveness, customers do not have to pay.
P.139

Dang Vo
www.behance.net/DangVoDesign

Dang vo is a young graphic designer and illustrator based in Montreal. He graduated from Ulam University. He is looking forward to becoming a multidisciplinary designer.
P.198-199

Distil Studio
www.distilstudio.co.uk

Distil Studio is a clear-thinking branding agency.

They filter out unnecessary fluff and clutter to reveal compelling brand identities, packaging, and visual communication. They are here to distil creative challenges into original, full-strength ideas.
P.075

Eskimo design studio
eskimodesign.ru/en

Eskimo design studio is one of the leading design agencies in Russia. Their core abilities cover various creative fields including logo design, brand identity, and web design. They work worldwide elaborating subtle ideas with strong visual character for a diverse range of brands and clients.
P.184-185

Espacioblanco
www.espacioblanco.de

Espacioblanco is a German/Brasilian creative team founded by Dennis Sommer and Talita Santos. They work in the field of visual identity.
P.120-121

Estudio Yeyé
estudioyeye.com

Estudio Yeyé, a branding and design studio with international presence, seeks day-to-day the most beautiful act of nature — design. They offer a wide range of services, such as graphic design, illustration, interior design, innovative branding, and marketing solutions. Their goal is simple — to provide their clients with relevant, innovative, and high-quality work to help their business grow.
P.226-227

Fable&Co.
fableco.uk

Fable&Co. is a branding, design, and digital creative agency. They bring new brands to life and breathe new life into the existing ones. Their artistry encompasses both print and digital applications across a variety of industry sectors. From their design studio in Brighton and Hove, they consistently deliver the exceptionally high quality works for which they are becoming renowned.
P.086-087

Fang Yi Chu
www.behance.net/FangYiChu

Fang Yi Chu, born in Taipei, Taiwan, is a graphic and visual designer currently living in New York. She has a Bachelor's degree of Civil Engineering and graduated from the M.S. Communication Design of

Pratt Institute. The diverse background turns her into a multidisciplinary designer who is adept at conducting structural thinking and bringing out the value of various brands through her aesthetics.

P.182-183

Futura
byfutura.com

Futura, founded in 2008, is a multidisciplinary design agency based in Mexico. They are experts in their field, and their scope goes beyond what is visual. They have experience in brand business strategy and creation. Futura understands what is best for the clients and their projects. The success of the brands is their main goal.

P.200-201

Gen Design Studio
jackmorgan.com

Founded in 2005, Gen Design Studio is a company dedicated to the development of graphic, product, environmental, illustration, UX, and web design. They see design as a corporative emancipation discipline. With this strategic perspective in mind, they are constantly aware of the world around them, predicting future scenarios and defining lines of sustainable action.

P.232

Geometry Global
www.geometry.com

Geometry Global is the world's largest and most international award-winning design agency. They dedicate to the programs that will bring change and inspiration to people. Geometry Global excels in digital, shopper, experiential, relationship, promotional, and trade marketing. Geometry Global is a WPP company.

P.176-177

Grantipo
www.grantipo.com

Grantipo is a design studio based in Madrid, specializing in branding and packaging. Grantipo offers solutions based on simple, functional, and coherent design. They are constantly learning and experimenting with creativity.

P.151

Grupo Ingenio
grupoingenio.com

Grupo Ingenio is a creative studio working on visual communication projects for print and new media.

P.138

Hani Douaji
www.hanidouaji.com

Hani Douaji is an award-winning conceptual and visual designer from Lebanon, who was born and grew up in Damascus-Syria. He holds a BA degree in Fine arts and Visual Communications from Damascus University and a MA degree in Graphic Design from the University of Central Lancashire, School of Arts, Design and Performance.

P.062-063

Hyela Lee
hyelalee.com

Hyela Lee is a graphic designer based in NYC. She has been awarded by AIGA, How Magazine, ADDY, IDA, Communication art, and Adobe.

P.058

IF BAGS
www.ifbags.it

IF BAGS is a backpack company in the heart of Milan. It is the perfect place to create customized backpack.

P.072

Iglöo Creativo
www.igloocreativo.com

Founded in 2012, Iglöo Creativo is a design studio whose services reach all ranges of branding spectrum. They specializes in the realization of corporate visual identity as well as packaging, posters, publishing, and web design. They believe that in a conceptual, simple, and creative design, what is really important resides in the minimal details.

P.060-061

I-Media Creative Bureau
www.imedia.kg

I-Media Creative Bureau is a design agency based in Bishkek, Kyrgyzstan. They focus on graphic design, branding, editorial design, and advertising.

P.108-109

Irene Acosta
www.irelexx.com

Irene Acosta is a Venezuelan-born graphic designer based in Los Angeles, CA. She loves exploring vibrant but minimalist concepts, and has a deep passion for urban art and culture. Irene graduated from Graphic Design and Interactive Media and Web Design in 2015 at Lindenwood University.

P.052-053

Jacob Bang
www.jacobbang.com

Jacob Bang is a Swedish graphic designer currently based in San Francisco. He is a graduate of Academy of Art University in San Francisco and Berghs School of Communication in Stockholm. He works across disciplines such as packaging, identity, motion, and digital design.

P.104-105

Jens Marklund
www.Jens.work

Jens Marklund is a New York based conceptual graphic designer, who works with a wide variety of mediums, materials, and processes. His designs come from a hand-on process aiming towards a conceptual and experimental outcome. Having grown up in Sweden, his Scandinavian roots help him keep a minimal aesthetic.

P.043

Jiani Lu
www.lujiani.com

Jiani Lu, born in China, is a multidisciplinary graphic designer and photographer currently living and working in Toronto, Canada. She received her Bachelor degree at York University, Sheridan College. Her work takes on a simplistic, minimalistic, and understated tone that she transfers across her print design, branding, and packaging design. She has won several awards including Adobe, D&AD, AIGA, The Dieline, and so on.

P.066-067

Joe Haddad
www.behance.net/haddaddesign

Joe Haddad is a multidisciplinary designer based in NYC. He currently studies at the School of Visual Arts to earn his BFA in Graphic Design. He has worked for several design agencies including Deutsch, Mother NY, GrandArmy, and most recently Chermayeff, Geismar, and Haviv.

P.166-167

Johannes Schulz
www.johannes-schulz.com

Johannes Schulz is a designer based in Hamburg, Germany who works across the fields of packaging, corporate, and 3D-design. He is fascinated by the beauty of simple yet eye-catching design solutions for every kind of business and clients from all over the world.

P.122-123

Kasperi Salovaara

www.behance.net/kasperisalovaara

Kasperi Salovaara is a graphic and packaging designer from Helsinki, Finland. He has finished his study at the Lahti Institute of Design and is now working at BOND Creative Agency.

P.142-143

Keiko Akatsuka

www.keikored.tv

Keiko Akatsuka is a freelance designer based in Japan.

P.059

Kevin Harald Campean

www.behance.net/HaraldKevin

Kevin Harald Campean is a freelance visual artist based in Budapest, Hungary. He specializes in creating outstanding brands through conceptual design, photography, and art direction.

P.190-191

kissmiklos

kissmiklos.com

Miklós Kiss's works incorporate various facets of architecture, fine art, design, and graphic design. A strong artistic approach and outstanding aesthetic quality characterize his art. His fine art pieces are just as significant as his distinctive style in corporate identity and graphic designs.

P.136-137, 206-207

Km Creative

www.kmcreative.gr

Km Creative is an agency dedicated to graphic design and visual communication projects including identity, print advertisement, website graphic, and packaging for various clients. Km creative offers creative and innovative solutions that are strongly associated with emotional responsiveness and spark the imagination.

P.032-033

Lam Cheung

www.behance.net/zeromultiply

Lam Cheung is a Hong Kong based graphic designer. Curiosity and passion drive him in delving into various creativity sectors. For him, good design is rendered through speculative observation and relevant details. Lam set up his own studio — invisible in 2013.

P.038-039, 042, 214, 218

Lap Yeung Hau

www.behance.net/jordanhau

Currently studying illustration at the University of Michigan, Lap Yeung Hau seeks to bring surprises to everyday life by injecting her illustrations of mundane daily happenings into lifestyle products. Having launched hand-illustrated stationery brands Randomese and Tatata, and social media character Avocadude on Wechat, Hau is now diving into the commercial corporations and the industrial world.

P.010-011

Latona Marketing Inc.

www.latona-m.com

Founded by Kazuaki Kawahara in Japan in 2008, Latona Marketing Inc. is a global award-winning graphic design studio adept at marketing. Their main business areas are graphic design, branding, packaging, product design, website, corporate design, corporate identity, advertising, and marketing.

P.022-023

Lavernia & Cienfuegos

www.lavernia-cienfuegos.com

Lavernia & Cienfuegos is a multidisciplinary design studio based in Valencia, Spain. They are a global agency specializing in graphic, product, and packaging design. To them, great design has an emotional component that connects with the consumer.

P.216

Leonie Werts

www.leoniewerts.nl

Leonie Werts is an enthusiastic designer from the Netherlands with a passion for geometry and details. Her work takes on a form of simplicity. Leonie is always searching for different ways to attract people's attention through design. Her inspiration comes from patterns and structures.

P.091

Lilkudley

www.lilkudley.cz

Lilkudley, a.k.a. Petr Kudláček, is a freelance graphic designer and art director based in Prague, Czech Republic. He loves working on printed and digital projects. For each project, he always wants to have under control the whole creative process, from custom illustration, logotype, typography, to layout and animation concept of responsive website.

P.054-055

Luminous Design Group

luminous.gr

Luminous Design Group is an award-winning creative agency providing integrated design and branding services since 2002. They are a team of designers with expertise in different skills, including software engineers and business developers, providing new alternative solutions beyond the predictable and traditional.

P.028

MADE

made-studio.ru

MADE is a team of young professionals in the area of corporate branding and website development based in Russia. For each task, they use a customized approach and maintain full control over the entire process, from sketches to development.

P.073

Maria Romero, Laura de Miguel, Cristian Varela

www.behance.net/MOMMA (Maria Romero)
www.behance.net/laurademig8e57 (Laura de Miguel)
www.behance.net/cristianvarela (Cristian Varela)

Maria Romero, Laura de Miguel, Cristian Varela are graduate students majoring in Packaging Design at ELISAVA University in Barcelona, Spain.

P.036-037

María López Benítez

www.behance.net/mariababia

María López Benítez is an industrial designer from Barcelona, Spain. After her study in Industrial Design at the University of Seville, she pursues her Master degree in Packaging Design at ELISAVA University.

P.168

Marios Karystios

www.karystios.com

Marios Karystios studied graphic design at the Technological Institution of Athens. He has worked as an art director in advertising for some established brands such as WIND and LG. In 2012 he set up his own studio in Cyprus countryside from where he works on diverse design projects. His work has been awarded and published in newspapers, magazines, and numerous blogs worldwide including Wallpaper*, Elle, Dezeen, and designboom, etc.

P.117, 156-157

Matthieu Jeanson
www.behance.net/matthieujeanson

Matthieu Jeanson is a graphic designer based in Montreal. His work reveals above all the sense of the project, through the exploration of shapes, colors, and typography.
P.116

Menta
www.estudiomenta.mx

Menta, founded by Laura Méndez in 2008, is a branding and packaging studio based in Guadalajar, Mexico. They believe in the *Simplicity of Allure* and deliver effective brand identities that balance classic and contemporary aesthetics. Their work builds meaningful human connections and stirs up beautiful experiences.
P.210-211

Metaklinika
www.metaklinika.com

Established in 2008 by Nenad Trifunović and Lazar Bodroža, Metaklinika is an independent design studio from Belgrade, which is active in the fields of creative services, initiatives, and production. Today, Metaklinika has ten permanently dedicated creative professionals, as well as a wide network of collaborators throughout the region.
P.078-079

mousegraphics
www.mousegraphics.eu

With a creative team consisting of ten designers, an illustrator, a photographer, a creative strategist, and an office manager, mousegraphics has considerable expertise in packaging design. A plethora of prestigious international awards and publications have placed mousegraphics among the most interesting, trustworthy, and better technologically equipped agencies. Their works range from logotype, packaging, and development of promotional material, etc. Mousegraphics is a member of EDEE, the Greek association of advertising and communication agencies, and Design Lobby.
P.056-057, 074, 128, 230-231

Mun Joo Jane
www.behance.net/munjoo

Mun Joo Jane is a multidisciplinary graphic designer who persistently searches for new ways of defining how people design and what design should look like. Most of his works display his interest in understanding human behaviors and the formation of cultures.
P.169

MUZIK
www.muzikeyewear.com

Established in 2013 with great focus on design and practicality, MUZIK maintains the competitive advantage through creating two brands, each with unique style of their own — MUZIK and STEALER. They utilize the best raw materials along with the highest techniques of craftsmen from South Korea and France for the creation of high-quality eyewear.
P.220-221

NANOIN Design
www.nanoin.cn

NANOIN Design is a design studio established in 2009 by Gao Fenglin, a doctor of design science. The name "NANOIN" is derived from a unique innovation method in his continuous research — Micro-design philosophy. NANOIN has won international design awards including Red Dot and iF Design Award.
P.101

Natasha Frolova, Louise Olofsson
www.behance.net/db_natasha846c (Natasha Frolova)
www.behance.net/LouiseOlofsson (Louise Olofsson)

Natasha Frolova and Louise Olofsson are designers from Sweden who have recently graduated as a packaging designer.
P.016-017

nendo
www.nendo.jp

Established by Oki Sato in 2002, nendo is a multi-disciplinary design agency based in Japan. Driven by the desire to create small "!" moments in daily life and overflowed with imagination, nendo has created a prolific amount of projects in architecture, interior, product, and graphic design. Their works have been displayed in various museums and galleries.
P.012-013, 020-021, 044-045, 110-111, 112-113

Nikita Konkin
www.behance.net/nikitos

Nikita Konkin is a graphic designer from Moscow, Russia. He enjoys observing the world to get new emotions and knowledge, and share them with others through design. He has participated in several design contest and won some awards including Pentawards.
P.046-047

Not Available Design
www.notavailable.hk

Established in 2009 by Kit Cheuk, Jerry Chow, and Sung Billy, Not Available Design (NA) is a multi-disciplinary studio based in Hong Kong, focusing on branding, design strategy, art direction, interactive media, and space design. They have won quite a few awards worldwide. Their philosophy is NOT following a set routine AVAILABLE in the market.
P.080-081

Nora Kaszanyi
www.behance.net/norakaszanyi

Nora Kaszanyi fell in love with graphic design in a young age. She is currently studying at the Moholy-Nagy University of Art and Design. She always strives to develop some designs which provide users with friendly experience. For her, design is a challenge, in which the efficient work is equally as vital as the mental contemplation.
P.228-229

Olivia Chan
www.behance.net/oliviachnn

Olivia Chan is a Montreal-based graphic designer who has recently graduated from at Dawson College. Olivia is currently working as a freelancer. She is mainly recognized by her minimalist aesthetics and bold energy conveyed in her work.
P.219

Ono and Associates Inc.
www.onoaa.com

Ono and Associates Inc. is an architecture and design office established by Takeshi Ono and Ayako Ono based in Toyota, Japan. They set a global vision and work broadly in various domains, such as architectural design, graphic design, and product design.
P.048

Ragini Sahai
www.raginisahai.com

Ragini Sahai is a graphic designer born in Gothenburg, Sweden who currently resides and works in Los Angeles. She loves to solve problems, communicate ideas, and render the abstract into the aesthetic. She has a wealth of experience in all areas of design, including brand identity, packaging, advertising, motion graphics, interactive, and web design.

P.049

Rasmus Erixon
www.rasmuserixon.com

Rasmus Erixon is a Swedish designer who is addicted to smart packaging and minimalistic style with eye-catching typography. On top of that he is a big fan of his camera. He has recently finished his studies at Brobygrafiska, College of Cross Media. Rasmus now works as a graphic designer in Stockholm, Sweden.

P.026-027, 064-065, 146-147, 194-195

Reverse Innovation
www.reverseinnovation.com

Reverse Innovation, with offices in Milan and Amsterdam, is a brand communication and product design agency with a strong commitment to innovation. A multidisciplinary team, careful research, and precise design enable Reverse Innovation to help clients develop and assert their brand.

P.130-131

Rice Creative
www.rice-creative.com

Established in 2011 by Chí-An De Leo and Joshua Breidenbach, Rice Creative is a branding and creative studio based in Vietnam. They are a multi-cultural team with perspective and precise vision. They strive to bring high value to bold brands through singular ideas with conviction and acclaimed craft. Rice Creative has gained an international reputation for developing powerful startup brands such as Marou Chocolate as well as edgy, award-winning work for international giants like Coca-cola and UNICEF.

P.132-133, 148-149, 212-213

Robot Food
www.robot-food.com

Robot Food is a group of passionate optimists, rebels, artists, and writers who believe that they can shape the future. Enjoying what they do is their recipe for producing great work.

P.154-155

Rong
www.rong-design.com

Rong is a creative agency located in Shanghai. They work through various fields and want to

observe and understand life differently by breaking established boundaries and barriers.

P.009, 152-153

Savvy Studio
savvy-studio.net

Savvy Studio is a branding and architecture design practice based in New York, Mexico City, and Monterrey. Savvy's expertise involves working around the globe on different ventures including boutique hotels, restaurants, retail spaces, art galleries, and museums.

P.217

Science Agency
www.welovescience.pro

Science Agency is a small and witty agency working in corporate and consumer branding. Their team love science, common sense, and problem solving. There is always a deep and well-balanced reasoning behind their works, as they design based on data. They do not believe in inspiration and have an unbreakable taboo against accidental results.

P.174-175

Shanghai Version Design
www.version-sh.cn

Shanghai Version Design, founded by Mori Zou, is an independent design studio based in Shanghai, China, focusing on branding, identity, image, and culture.

P.186-187

Shao
www.designbyshao.com

Shao is a multidisciplinary design firm based in Manila, Philippines. They create brands, stories, experiences, and content across all platforms. They work with experience and passion that are rooted in research and strategic thinking to create conceptually driven intelligent work.

P.204-205

Sofia Villarreal Samperio, Cynthia Fernandez Arellano
www.behance.net/svillarreals (Sofia Villarreal Samperio)
www.behance.net/cindyfdza (Cynthia Fernandez Arellano)

Sofia Villarreal Samperio and Cynthia Fernandez Arellano currently study at the Universidad de Monterrey in Monterrey, Mexico.

P.202-203

Sophie Giraldeau
www.behance.net/sophie_gir18d0

Sophie Giraldeau is a third-year student in Graphic Design at the Université du Québec à Montréal (UQAM) and a junior graphic designer. She is passionate about all elements in graphic design ranging from typography, packaging, branding, and illustration.

P.124-125

Spencer Hill
www.spencerhill.me

Spencer Hill is an art director, designer, and developer. Located in Brooklyn, Spencer completed his creative journey from Pratt Institute in 2015 with a degree in Communications Design. His experience covers product design, branding, art direction, web design, and 3D graphics.

P.029

Stas Neretin
www.behance.net/1iz25

Stas Neretin was born in Voronezh, Russia. He graduated from the British Higher School of Design in Moscow in 2015 with a degree in architecture and design. After graduation, he moved to Moscow and worked as a graphic designer in many agencies. He appreciates the humor and some cool stuff derived from a joke or funny contrivance. He enjoys design that evokes emotions.

P.114-115

Stelios Ypsilantis
www.steliosyps.com

Stelios Ypsilantis is an independent graphic designer from Kozani, who works in Greece and the United Kingdom. His creative practice focuses on clear communication with an emphasis on typography, editorial, conceptual design, packaging, and illustration.

P.118-119

Studio Sonda
sonda.hr

Studio Sonda is an independent, award-winning creative studio for design and communications. They excel in visual communications, brand conception and development, promotional strategies and campaigns, packaging design, and web design.

P.098-099

Talia Douaidy
www.taliadouaidy.com

Talia Douaidy is a New York based packaging designer who has acquired her Master degree in Packaging Design at Pratt Institute. She was born and raised in a Lebanese family in Paris. The eclectic background and cultural exposure inspire her on design. Her latest work involves the use of a wide range of materials and innovative production techniques.
P.014-015

Thibault Magni
www.thibaultmagni.fr

Thibault Magni is a student majoring in Graphic Design at the Université du Québec à Montréal (UQAM) in Canada. The most important element in his work is that the design must be accessible for everybody and communicates with the largest number of people.
P.140-141

Toby Ng Design
www.toby-ng.co

Graduated from Central St. Martins, London in Graphic Design, Toby Ng practiced his craft in London, Singapore, and Hong Kong, prior to starting his own design firm in Hong Kong in 2014. Specializing in graphic design and brand identity, Ng rigorously tackles design challenges with wit and aesthetically meaningful communications. Ng is winner of numerous awards including Red Dot, D&AD, etc. He was named in 2014 as one of the top 40 design talents under the age of 40 in Asia.
P.034-035

Tough Slate Design
toughslatedesign.com

Tough Slate Design is an independent, award-winning advertising and branding agency based in Kyiv, Ukraine. They are experts in advertising, branding, packaging, and graphic design with a working experience of over 40 years. They have received more than 200 awards on various advertising festivals.
P.093, 162-163, 164-165

Tres Tipos Gráficos
www.trestiposgraficos.com

Tres Tipos Gráficos, a dedicated and bold design agency, delivers customized and durable solutions for visual identity, books, packaging, and digital media. They provides solutions that are both efficient and lasting. They focus on areas that range from art direction to design, printing process and development.
P.178-179

TRIANGLE-STUDIO
www.triangle-studio.co.kr

TRIANGLE-STUDIO is an art and graphic design group specializing in branding, editorial design, and illustration. They create a brand design based on rational strategy and emotional harmony.
P.215

Tsan Yu Yin
www.behance.net/tyyin

Tsan Yu Yin is an independent graphic designer in the fields of typography, packaging, and curating. His work has been extensively featured in international design awards and publications. He loves rollercoaster.
P.040-041

twelvemonthly
www.12monthly.com

Twelvemonthly is a studio with creative people who spread collective thoughts to random audiences. They visualize their thoughts every month. They strive not only to inspire but also to encourage everyone to participate and add serial creativity.
P.070-071

Veej
www.behance.net/veej75

Veej is the studio name of a designer and 3D illustrator, Vijay Ram. Located in the Midlands city of Leicester, he has worked in the design industry for 16 years and currently works as a freelancer. He has been a designer at bigdog 8 years.
P.150

Velimir Andrejevic, Sofija Gavrilovic
boomerangstories.com

Velimir Andrejevic, a graphic designer, and Sofija Gavrilovic, a painter, have merged their expertise and artistic experience and set up the brand The Returning of the Boomerang. The boomerang acts as an artistic object (product) which combines traditional craft and new artistic practice.
P.068-069

Vinicius Hideki
viniciushideki.com

Vinicius Hideki was a student majoring in Product Design and currently works as a graphic designer. He is a city boy, art lover, bad cooker, and the worst writer. He believes in a minimal design which is practical and clean.
P.050-051

Zoo Studio
zoo.ad

Zoo is more than a graphic and multi-media agency. They create their global communication projects by working with clear concepts, simple graphics, and good ideas.
P.090, 102-103

Zsofi Ujhelyi
zsofiujhelyi.com

Zsofi Ujhelyi is an industrial product designer with Hungarian origin currently based in the Netherlands. She is skilled in packaging and window display. Her long-term aim is to reframe what and how things are designed in the next decade. She is also an avid cyclist and seeker of the divine.
P.082

Zunder
www.zunder.wtf

Zunder solves communication problems. They create relevance between the message and the recipient. They try every day to make the world a slightly better place.
P.096-097

Zup Design
www.zup.it

Founded in 2001, Zup Design is an Italy-based creative studio focusing on communication, industrial design, multimedia installations, exhibition spaces, and interior design. Zup approaches each issue from a global and comprehensive point-of-view, in the belief that the design is no longer to distinguish between graphic, industrial, and architectural design, but to engage instead with language, volumes, shapes, and images. Zup's work has been recognized with numerous awards, including Red Dot, European Design Awards, and so on.
P.024-025, 076-077

ACKNOWLEDGEMENTS

We would like to thank all of the designers involved for granting us permission to publish their works, as well as all of the photographers who have generously allowed us to use their images. We are also very grateful to many other people whose names do not appear in the credits but who made specific contributions and provided support. Without these people, we would not have been able to share these beautiful works with readers around the world. Our editorial team includes editor Jessie Tan and book designer Wu Yanting, to whom we are truly grateful.